The Dinosaur Questions

THE PLACE OF THE DINOSAURS IN THE REPTILE CLASS				
Class	Subclass	Order	Suborder	Some Examples
REPTILIA	Anapsida	Cotylosauria	Diadectomorpha	*Diadectes*, borderline amphibians
			Captorhinomorpha	*Solenodonsaurus*, *Captorhinus*, *Hylonomus*
			Parareptilia	Mesosaurs, Millerettids, Pareiasaurids, *Scutosaurus*
		Chelonia	Pleurodira	*Stupendemys*
			Cryptodira	*Archelon*, turtles, tortoises
	Synapsida	Pelycosauria	Ophiacodontia	*Ophiacodon*, *Tetraceratops*
			Sphenacodontia	*Dimetrodon*, *Sphenacodon*
			Edaphosauria	*Edaphosaurus*, *Cotylorhynchus*
		Therapsida	Biarmosuchia	*Biarmosuchus*
			Dinocephalia	*Moshops*, *Titanosuchus*
			Dicynodontia	*Kannemeyeria*, *Kingoria*, *Lystrosaurus*
			Gorgonopsia	*Lycaenops*, *Arctognathus*
			Therocephalia	*Theriognathus*, *Hyenosaurus*
			Cynodontia	*Procynosuchus*, *Thrinaxodon*, *Kayentatherium*
	Lepidosauria (Diapsids)	Eosuchia	Younginiformes	*Youngina*, *Tangasaurus*
			Choristodera	*Champsosaurus*
			Thalattosauria	*Nectosauria*, *Thalattosaurus*
			Pleurosauria	*Pleurosaurus*
		Rhynchocephalia	Sphenodontida	Tuatura
		Squamata	Sauria	lizards, mosasaurs
			Serpentes	Snakes
	Archosauria (Diapsids)	Thecodontia	Pseudosuchia	Aëtosaurs, Rauisuchians, *Euparkeria*, *Ornithosuchus*
			Phytosauria	Phytosaurs, *Rutiodon*
		Crocodilia	Protosuchia	*Protosuchus*
			Mesosuchia	Teleosaurs, *Sebecus*
			Eusuchia	'modern' crocodiles, alligators, gavials, etc.
		Pterosauria	Rhamphorhynchoidea	*Dimorphodon*, *Rhamphorynchus*
			Pterodactyloidea	*Pteranodon*, *Quetzalcoatlus*
		Saurischia	Theropoda Sauropodomorpha	DINOSAURS
		Ornithischia	Cerapoda Thyreophora	
	Euryapsida	Sauropterygia	Placodontia	*Placodus*, *Cyamodus*
			Nothosauria	*Nothosaurus*, *Lamprosaurus*
			Plesiosauria	*Pliosaurus*, *Elasmosaurus*
		Ichthyopterygia	Ichthyosauria	*Ophthalmosaurus*, *Ichthyosaurus*, *Mixosaurus*

This is a simplified classification chart which omits problem areas such as the early diapsids (e.g. *Petrolacosaurus*, rhynchosaurs and *Tanystropheus*) which are not assigned to a particular suborder.

THE DINOSAUR QUESTIONS

A Quiz-book for the Enthusiast

IAN D. TYAS

PAUL WATKINS
STAMFORD

Typeset and Published by
Paul Watkins
18, Adelaide Street
Stamford
Lincolnshire, PE9 2EN.

ISBN

1 871615 60 7

Printed and bound by Woolnoughs of Irthlingborough

FOR GEORGE

CONTENTS

PREFACE

This Dinosaur Quiz Book has been designed for the dinosaur enthusiast of all ages. Over a thousand questions and answers have been arranged into thirty-four quizzes in approximate order of difficulty. The questions concentrate on factual knowledge as befits a 'quiz book', so you will find no 'Join-the-dots-osaurus' or 'help Iguanodon through the maze' here. For variety there has been included a minimum of questions on *other* prehistoric animals such as the dinosaur ancestors and contemporaries and mammals.

A difficulty which arises in writing any book about dinosaurs is that, although the subject of the book died out some sixty-five million years ago, our knowledge about them is both recent and rapidly changing. New discoveries are made and old facts are reinterpreted so constantly that any dinosaur book is out-of-date as soon as it is printed. Despite this, every effort has been made to ensure the accuracy of the information presented in the answers. I hope that any remaining errors do not spoil the pleasure and fun for which the book was intended.

In compiling the dinosaur classification chart reference was made to as much recent information as possible. Nevertheless, much of it remains controversial, and preference has often been given to traditional classifications where these clashed with recent theories. *Dilophosaurus*, for example, is traditionally a megalosaurid, but could it be a ceratosaurid or an overgrown coelophysid? Or should it belong to a family of its own? The 'protodinosaurs' are often attached to the prosauropods and the recent changes to the status of *Troödon* and *Scutellosaurus* exemplify the difficulties in presenting a classification of such a rapidly-changing subject.

Finally, I wish to thank those who have helped, especially Jane, who patiently tolerated the covering of the kitchen table with hundreds of books and papers; Graham who kindly agreed to read through the text; Marion who did the typesetting and Shaun who first suggested the idea. Thanks also, for general enthusiasm, encouragement and interest, to: Amanda, Amelia, Andrew, Andy, Anna, Bozz, Carol, Carolyn, Chris, Cloe, Craig, Daniel, Dave, David, Dawn, Declan, Diane, Donald, Doreen, Eleanor, Faye, Gemma, George, Jackie, James, Jerome, Jim, Joanne, John, Jösef, Joseph, Kirsteen, Laura, Liz, Lyndsey, Margaret, Mark, Mike, Mother, Natalie, Patrick, Paul, Peter, Philip, Ray, Richard, Robert, Rosamund, Sam, Sara, Stephen, Thomas, Tim, Tom, Victoria, William and Witford.

Saltburn-by-the-Sea
June 1993

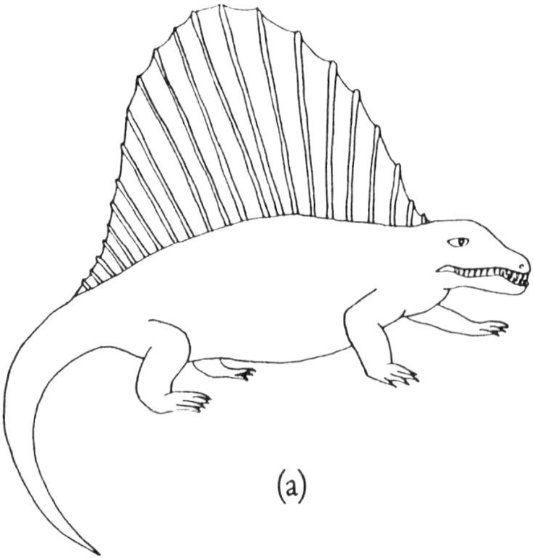

(a)

(b)

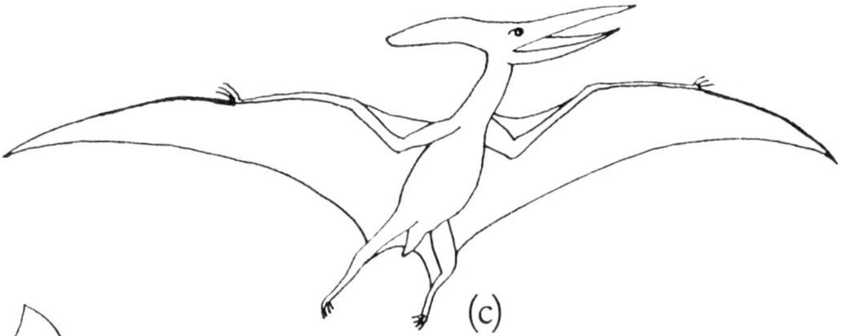

(c)

(d)

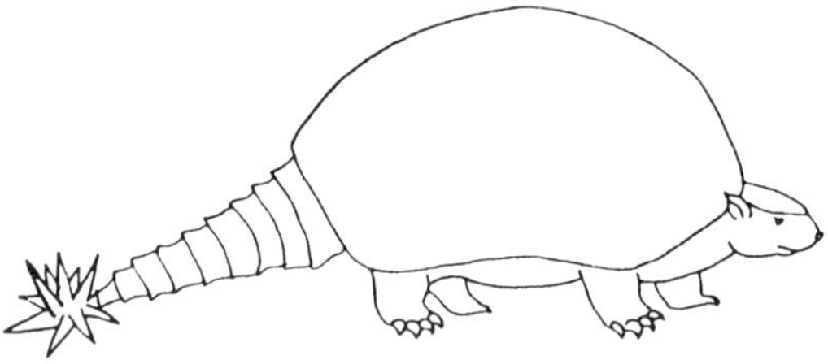

(e)

Quiz 1

1. The word 'dinosaur' comes from the Greek *deinos* and *sauros*. What does it mean?
2. Who first used the word 'dinosaur'?
3. In which year was the word 'dinosaur' first used?
4. Which, if any, of the animals pictured on the opposite page can properly be called dinosaurs?
5. In films and on television, cavemen are often shown fighting off dinosaurs. Why is this nonsense?
6. Identify this dinosaur:

7. Most of us have heard of *Brontosaurus*, but it should properly be called by the name it was first given. What name was it first given?
8. Of which long dinosaur is 'SUD COOP LID' an anagram?
9. 'All dinosaurs were stupid creatures with small brains.' True or false?
10. Which of the following three statements is correct?
    (a) Dinosaurs lived in the sea and on land.
    (b) Dinosaurs lived only on the land.
    (c) Dinosaurs lived in the sea, on the land and in the air.
11. How long ago did the dinosaurs become extinct?
12. In which geological era did the dinosaurs live?
13. Because *Brachiosaurus* had its nostrils at the top of its head, it was once believed that it stood on the bottom of a lake and used its long neck as a snorkel. Why is this now regarded as impossible?

14. Which group of dinosaurs had distinctively-shaped crests on their heads?
15. At the beginning of the dinosaur era, the world had one large land mass. What do scientists call this supercontinent?
16. How old do scientists believe the world to be?
17. Which is the smallest known dinosaur?
18. In terms of posture, what is the difference between dinosaurs and modern reptiles like crocodiles?
19. Here are three dinosaur feet. One belongs to a sauropod, one a theropod and one an ornithopod. Which is which?

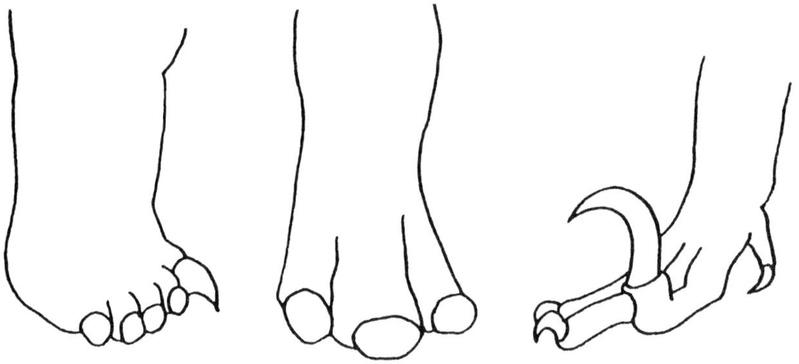

(a) (b) (c)

20. What does the name *Struthiomimus* mean?
21. Was *Iguanodon* a herbivore (vegetarian), a carnivore (meat-eating), or an omnivore (eating both types of food)?
22. What is the name given to the science which includes the studying of dinosaurs and other prehistoric animals?
23. Dinosaurs are classified into two groups called orders, the essential difference between them being their type of hip joints. What are these two orders called?
24. Some scientists claim that not all dinosaurs died out – they evolved into a class of very well-known animals. What are these well-known animals?
25. Where is the largest mounted dinosaur skeleton in the world on public display and which dinosaur is it?
26. Which dinosaur had plates standing up along its back?

27. 'All dinosaurs laid eggs like birds, rather than producing live young like mammals.' True or false?
28. Of which horned dinosaur is 'RICE STOP RAT' an anagram?
29. Which could move the fastest – *Iguanodon* or *Hypsilophodon*?
30. How long was a typical *Brachiosaurus*: 3,13, 23, 33 or 43 metres?

Quiz 2

1. Why should the names of dinosaurs really be written in italics?
2. Which came first, the dinosaurs, the mammals or did both evolve more or less simultaneously?
3. For how many millions of years did the dinosaurs rule the Earth?
4. Which is the longest known dinosaur?
5. Which dinosaur is usually described as being the first dinosaur to be discovered (in 1822)?
6. (a) Who found and (b) who named this first dinosaur in 1825?
7. Identify this dinosaur:

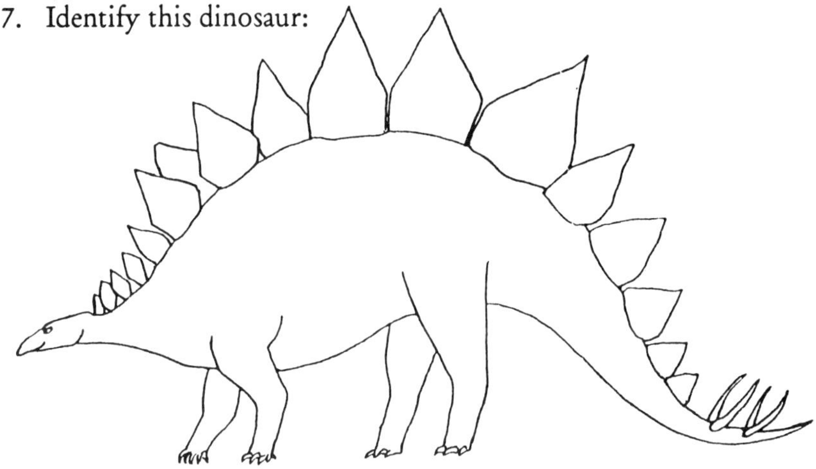

8. Which group of dinosaurs is called the 'horned dinosaurs'? Is it the Iguanodontidae, the Ceratopsidae, the Ankylosauridae, the Pachycephalosauridae or the Hadrosauridae?
9. It was once believed that dinosaurs were cold-blooded reptiles. What does 'cold-blooded' mean?
10. What does 'warm-blooded' mean?
11. Which four-legged dinosaur had longer legs at the front than at the back?
12. 'Dinosaurs died out because they couldn't stand the competition with mammals.' True or false?

13. Of which meat-eating dinosaur is 'A NASTY SOUR URN' an anagram?
14. What kind of rock is most likely to contain fossils?
15. What do the terms (a) *saurischian* and (b) *ornithischian* mean?
16. If you describe something in the world today as 'a dinosaur' (for example a government department!) what is meant by it, and why is it wrong?
17. What do the two terms bipedal and quadrupedal mean?
18. At the end of which geological period did the dinosaurs die out?
19. How many millions of years ago do scientists believe life began to evolve?
20. Of the two dinosaur orders Saurischia and Ornithischia, one is completely vegetarian while the other is half herbivore, half carnivore. Which is which?
21. Who could run fastest: *Apatosaurus* or a human being?
22. What is a 'living fossil'?
23. What kind of animal was *Brontotherium*?
24. Which dinosaur was the 'thick-headed reptile'?
25. 'All dinosaurs were enormous.' True or false?
26. In the middle of the dinosaur era, the supercontinent of Pangaea had split into two large continents. What do we call these 2 large continents?
27. What is the distinctive feature of *Spinosaurus*?
28. Why is *Protoceratops* often depicted hatching out of eggs?
29. Which is the odd-one-out and why: *Allosaurus, Megalosaurus, Deinonychus, Parasaurolophus* or *Ceratosaurus*?
30. Correct this spelling mistake: STREWTH I OWE MEEMUS.

Quiz 3

1. The second dinosaur to be discovered, in 1824, became the first to be named. (a) Which dinosaur was it? And (b) who gave it its scientific description?
2. What were the 'thecodonts'?
3. 'Dinosaurs had two brains – one in their heads and one in their bellies.' True or false?
4. Which group of dinosaurs is known as the 'armoured dinosaurs'?

5. Throughout geological time the Earth's land masses have moved about. What is this phenomenon called?
6. What are the scientific terms for (a) cold-blooded and (b) warm-blooded?
7. Identify this dinosaur:

8. What does the name 'archosaur' mean?
9. *Tyrannosaurus* had grotesquely small and seemingly useless arms. What was their probable use?
10. Which is the biggest and heaviest known dinosaur?
11. Of which well-known dinosaur is 'GUANO NOID' an anagram?
12. In 1982 William Walker discovered a new dinosaur in a Surrey clay pit. It became nicknamed 'Claws'. What name was given to it later?
13. 'Claws' is believed to have been piscivorous. What is a piscivore?
14. What is the *first* essential requirement for fossilisation?
15. '*Dimetrodon* was one of the earliest dinosaurs.' True or false?
16. *Diplodocus* was typically as long as approximately how many double-decker buses?
17. Correct this spelling mistake: SHTEGGIO SAW RUSK.
18. What was the most distinctive feature of *Deinonychus*?
19. Most dinosaurs are referred to by one name. Is this the name of the genus or the species?
20. In which American state is there a town called 'Dinosaur'?
21. Who was the 'father' of palaeontology?
22. What kind of animal was *Eryops*?
23. What does the name *Oviraptor* mean?
24. *Ichthyosaurus* was a reptile that lived at the same time as dinosaurs. Why is *Ichthyosaurus* not a dinosaur?

25. What was found in April, 1878 in a coal mine at Bernissart, Belgium?
26. What are the two most obvious differences between the two heavily-armoured dinosaurs *Nodosaurus* and *Ankylosaurus*?
27. All dinosaurs belong to either the bird-hip (*ornithischian*) or the reptile-hip (*saurischian*) orders. Below are illustrated these two types of hip joint. Which is which?

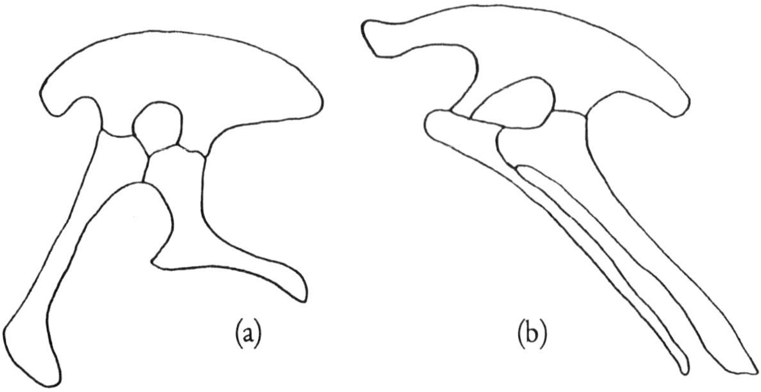

28. What were the 'theropods'?
29. In which country are most of the new dinosaurs being found today?
30. What was unusual about *Heterodontosaurus* - as is implied by the meaning of its name?

Quiz 4

1. When Richard Owen first used the word 'dinosaur', in 1841, which were the three genera already known?
2. Where was the Tethys Sea?
3. Which is the fastest known dinosaur?
4. *Brachiosaurus* was as heavy as approximately how many elephants?
5. When scientists study the bones of dinosaurs they try to put them together in the right places but sometimes they get it wrong. What was the most important early mistake with *Iguanodon*?
6. 'All dinosaurs were ferocious carnivores.' True or false?
7. It was a widely held belief that dinosaurs had two brains. This is certainly not true, but on what evidence could such a belief arise?
8. What is 'extinction'?

9. What kind of animal was *Ichthyostega*? What was special about it?
10. Identify this dinosaur:

11. Correct this spelling mistake: TIRANOSAURUS WRECKS.
12. Which could move faster: *Stegosaurus* or *Struthiomimus*?
13. In classifying dinosaurs and other animals, individuals are known as specimens while a group of similar specimens are known as a species. Because we know so little about dinosaurs, it is often not possible to distinguish between individual species, so we usually refer to dinosaurs by their generic names. What is the next level up from genus into which are grouped similar genera?
14. What was the distinctive feature of *Parasaurolophus*?
15. What is the name given to the order of flying reptiles?
16. Dinosaurs lived in the Mesozoic era. What does 'Mesozoic' mean?
17. What are the three geological periods which make up the Mesozoic era?
18. Which is the odd-one-out: *Dimetrodon, Plesiosaurus, Mosasaurus, Megalosaurus* or *Archelon*?
19. Which dinosaur has been described as 'like a cantilever bridge'?
20. What was the name of the sculptor who first created life-size models of dinosaurs in the 1850s and whose models can still be seen in Sydenham Park in South East London?

21. Most scientists believe that the most accurate measure of intelligence lies in the ratio between brain size and body size. So, which dinosaur is usually regarded as being the most intelligent?
22. What is the connection between *Compsognathus*, a dinosaur, and *Bavarisaurus*, a lizard?
23. Of which dinosaur is 'RUST SAGO SUE' an anagram?
24. Why are there no dinosaur remains in coal?
25. Which group of dinosaurs were the 'bird mimic reptiles'?
26. Name the three bones of the pelvis.
27. Was *Triceratops* ornithischian or saurischian?
28. Was *Diplodocus* ornithischian or saurischian?
29. Which dinosaur was the 'parrot reptile'?
30. How long was a typical *Iguanodon*: 0.9m, 9m, 19m, 29m, or 39m?

Quiz 5

1. Who were the two great American rivals who searched for dinosaurs in the nineteenth century?
2. Was *Tyrannosaurus* ornithischian or saurischian?
3. Was *Iguanodon* ornithischian or saurischian?
4. Who first put forward the idea of continental drift?
5. Put these geological periods into the correct order: Carboniferous, Silurian, Precambrian, Ordovician, Devonian, Cambrian.
6. Of which large dinosaur is 'ACRA BRUSH IOUS' an anagram?
7. The first nodosaurid dinosaur to be discovered was *Hylaeosaurus* in 1833. What does its name mean?
8. The skull of *Tyrannosaurus* is massively reinforced in order to withstand collision. What does this tell us about its possible hunting method?
9. Correct this spelling mistake: TRY SIERRA TOPS.
10. Which dinosaur has the longest generic name?
11. Was *Tyrannosaurus* a herbivore, a carnivore or an omnivore?
12. What was the most striking feature of *Iguanodon*?
13. What does the name *Diplodocus* mean, and to what does it refer?
14. *Cetiosaurus* was a very early discovery (1809) but it wasn't recognised as a dinosaur until 1869, by T. H. Huxley. What was the reason for the delay?
15. What kind of an animal was *Dinichthys*?

16. Identify this dinosaur:

17. Is 'Brachiosauridae' the name of the species, the genus, the family, the infraclass or the subclass?
18. Why would *Stegosaurus* never have encountered *Psittacosaurus*?
19. Was *Stegosaurus* ornithischian or saurischian?
20. In 1915 the President of the United States ordered the 'Dinosaur National Monument' to be created. It is a working museum on the site of a dinosaur quarry. In which state does it lie?
21. In the 1920s the first dinosaur nest was found. Which kind of dinosaur was it and in which country was it?
22. What were the sauropodomorphs?
23. What was the most striking feature of *Stegosaurus*?
24. 'All dinosaurs walked on two legs.' True or false?
25. What were and are Anapsids, Synapsids, Euryapsids and Diapsids?
26. How many fingers did *Tyrannosaurus* have on its little arms?
27. How long was a typical *Allosaurus*? Was it 5, 11, 15, 18 or 21 metres?
28. What was the most striking feature of *Pachycephalosaurus*?
29. Which one of the following sciences is *irrelevant* to the study of dinosaurs: palaeontology, palaeohistology, palaeobotany, palaeography or palaeomagnetism?
30. According to the theory that intelligence is related to the brain size to body size ratio, which was the most stupid dinosaur?

Quiz 6

1. Put these geological eras into the correct order: Mesozoic, Cenozoic, Palaeozoic.
2. Put these geological periods into the correct order: Cretaceous, Permian, Triassic, Jurassic, Quaternary, Tertiary.
3. Was *Brachiosaurus* saurischian or ornithischian?
4. Identify this dinosaur which is drawn in a traditional pose:

5. In 1991 a European Community Summit was held at Maastricht in Holland. What happened at a chalk quarry there in 1770?
6. What are fossilised droppings called?
7. Of which carnivorous dinosaur is 'CHIN YOU SEND' an anagram?
8. What is a cladogram?
9. In continental drift, what is the name for the continent-sized 'sheets' upon which the continents ride?
10. Why was Bone Cabin Quarry so called? And where is it?

11. 'Some dinosaurs were no bigger than a chicken.' True or false?
12. Was *Triceratops* a herbivore, a carnivore or an omnivore?
13. Correct this spelling mistake: DY PLODDING KUSS.
14. Which dinosaur has the shortest generic name?
15. What was the typical length of the crest of an adult *Parasaurolophus*: nearly 1, 1.5, 2, 2.5 or nearly 3 metres?
16. Was *Parasaurolophus* a saurischian or an ornithischian dinosaur?
17. Name four major groups of animals that were all 'archosaurs', that is, 'ruling reptiles'.
18. Identify this pterosaur:

19. Why shouldn't we call the above pterosaur a flying dinosaur?
20. What kind of animal was *Uintatherium*?
21. The American dinosaur experts Othniel Marsh and Edward Cope were arch-rivals but they were friends until 1870. What ended their friendship?
22. In fossil formation, what is 'permineralisation'?
23. What was the probable purpose of the extraordinary thick skull of *Pachycephalosaurus*?
24. In the popular American television comedy series *Dinosaurs*, what is the family surname of Earl, Fran, Robbie, Charlene and Baby?
25. Which is the biggest animal ever?
26. Which dinosaur was the 'roofed reptile'?

27. Rocks are laid down in strata. What do we know about the strata (and consequently any fossils found in them) that are lower than the others?
28. Which is the odd-one-out: *Triceratops*, *Stegosaurus*, *Ankylosaurus*, *Styracosaurus* or *Ornithomimus*? And why?
29. It was once believed that *Hypsilophodon* was arboreal. What is 'arboreal'?
30. Which is the oldest of the known dinosaurs?

Quiz 7

1. Who is this famous dinosaur person?

2. 'The ornithischian (bird-hip) dinosaurs evolved into birds.' True or false?
3. 'The ornithopod (bird-foot) dinosaurs evolved into birds.' True or false?
4. Was *Dilophosaurus* an ornithischian or a saurischian dinosaur?
5. The first Canadian dinosaur remains were found in the Red Deer River Valley in the early 1870s. It was soon realised that there was only one practical way to collect them. What was it?
6. Which dinosaur laid the biggest known egg?
7. Of which crested dinosaur is 'OH PAUL, SARA POURS' an anagram?
8. What is the 'phanerozoic'?
9. What are the three possible processes which could lead to the formation of a stone fossil?

10. How many *species* of dinosaurs have so far been named? Is it 50-100, 100-200, 300-500, 500-900 or over 1000?
11. At the beginning of the dinosaur era, most dinosaur species were widespread throughout the world. Towards the end of the Cretaceous each species was confined to one or two continents. Why should this be so?
12. Identify this dinosaur, which is drawn in a very traditional pose:

13. Which dinosaur was the 'tyrant reptile king'?
14. Was *Ankylosaurus* an ornithischian or a saurischian dinosaur?
15. What have these dinosaurs got in common: *Triceratops, Styracosaurus, Chasmosaurus, Protoceratops* and *Centrosaurus*?
16. Which of these dinosaurs is the odd-one-out: *Chasmosaurus, Torosaurus, Stegoceras, Pentaceratops* or *Protoceratops*?
17. What kind of animal was *Diatryma*?
18. Correct this spelling mistake: A PATTIO DOORUS.
19. What is the most distinctive (indeed, the only known) feature of *Deinocheirus*?

20. What does the name *Dilophosaurus* mean?
21. Was *Dilophosaurus* a herbivore, a carnivore or an omnivore?
22. Who, in 1887, divided the dinosaurs into saurischians and ornithischians?
23. What are the two suborders into which the saurischian dinosaurs can be divided?
24. Here are a dolphin, a shark and an ichthyosaur. Which is which?

(a)

(b)

(c)

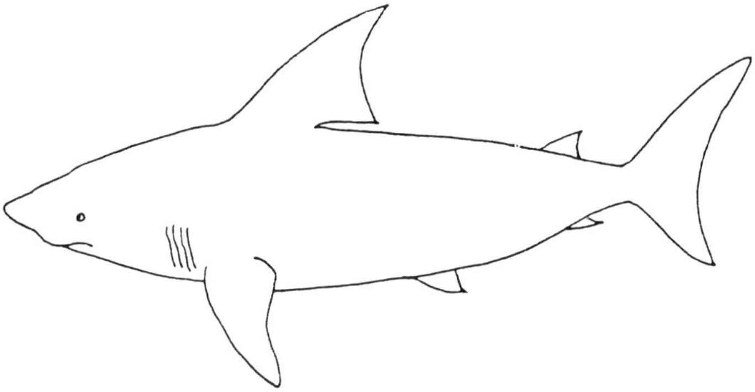

25. The forelimbs of *Allosaurus* are not as small as those of *Tyrannosaurus*. What would *Allosaurus* probably have used its arms for?
26. Which finger in the pterodactyls supported the wing?

27. '*Diplodocus* was not the longest dinosaur - *Seismosaurus* was even longer.' True or false?
28. In bipedal dinosaurs what was the usual function of having a long tail?
29. Some fossils suffer from pyrite disease or pyritic decay. What is pyrite disease?
30. Identify this dinosaur:

Quiz 8

1. During the Triassic period, the dinosaurs were beginning to evolve from a group of animals represented in the fossil record by *Proterosuchus, Erythrosuchus* and *Euparkeria.* What was the name of this group?
2. 'Dinosaurs were wiped out by early cavemen.' True or false?
3. Where a fossil dinosaur is found, for example in a cliff face, what is the 'overburden'?
4. The long-necked dinosaurs like *Diplodocus* and *Brachiosaurus* are all part of a large group within the order *Saurischia.* We call this large group an 'infraorder'. What name do we give to it?
5. Which dinosaur was the 'stiff, or fused, reptile'?
6. Sometimes large numbers of dinosaur remains are all found together. Which dinosaur holds the record for the greatest number of individuals found in one place, and where was it?

7. Identify this dinosaur:

8. Which two dinosaur families had no teeth?
9. Of which small dinosaur is 'POOR COT PASTER' an anagram?
10. What is the most distinctive feature of *Nodosaurus*?
11. In 1676 Dr Robert Plot published a book containing a picture now known to be that of a dinosaur bone. What did he believe it to be?
12. Some dinosaur experts claim that *Tyrannosaurus* was far too large and heavy to run after its prey for any length of time. How do these experts believe that *Tyrannosaurus* acquired its food?
13. Was *Diplodocus* a herbivore, a carnivore, or an omnivore?
14. What kind of animal was a phytosaur?
15. Why would *Euparkeria* never have met *Tyrannosaurus*?
16. Correct this spelling mistake: MEG HELLO SAW US.
17. Which country has the distinction of being the first to issue a postage stamp depicting a dinosaur?
18. Which was the first dinosaur to be depicted on a postage stamp?
19. 'All dinosaurs died out at the end of the Cretaceous.' True or false?

20. Identify this pterosaur:

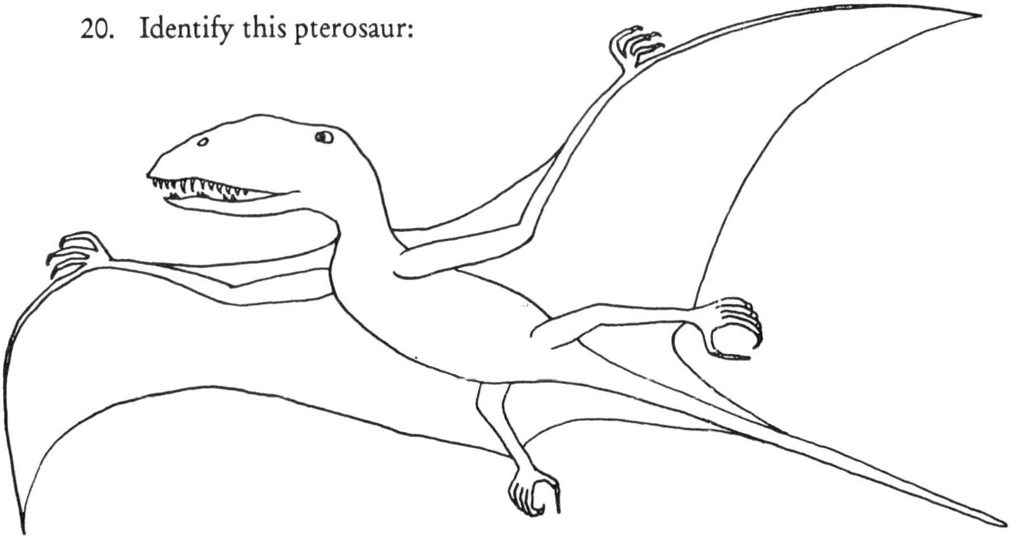

21. Identify this dinosaur from its description: It was a North American 4-footed dinosaur; nearly a third of its length was taken by its head; it was about 9 metres long and must have weighed 5-6 tonnes; it had three large horns on its face.
22. Which of the famous expeditions in the past was led by Roy Chapman Andrews?
23. Who is this famous dinosaur person?

24. Which of these dinosaurs weighed only 900 grams: *Stegosaurus*, *Iguanodon*, *Saltopus*, *Triceratops*, or *Anatosaurus*?
25. How many fingers and how many toes did *Struthiomimus* have?
26. Where would you find Tyrannosaurus Terrace, Brontosaurus Boulevard and Stegosaurus Freeway?
27. Why didn't absolutely enormous dinosaurs lay absolutely enormous eggs?
28. One dinosaur took armour plating to such an extreme that it even had bony eyelids. Which dinosaur was it?
29. What does the name *Hypsilophodon* mean?
30. What were the two possible purposes of *Iguanodon's* thumb spike?

Quiz 9

1. What was the most distinctive feature of *Ceratosaurus*?
2. When we talk about *Ankylosaurus* we are referring (strictly speaking) to the genus although in a looser sense we might mean the species. To what are we referring when we say 'Ankylosaur' or 'Ankylosaurid', and why is this name not written in italics?
3. Identify this dinosaur:

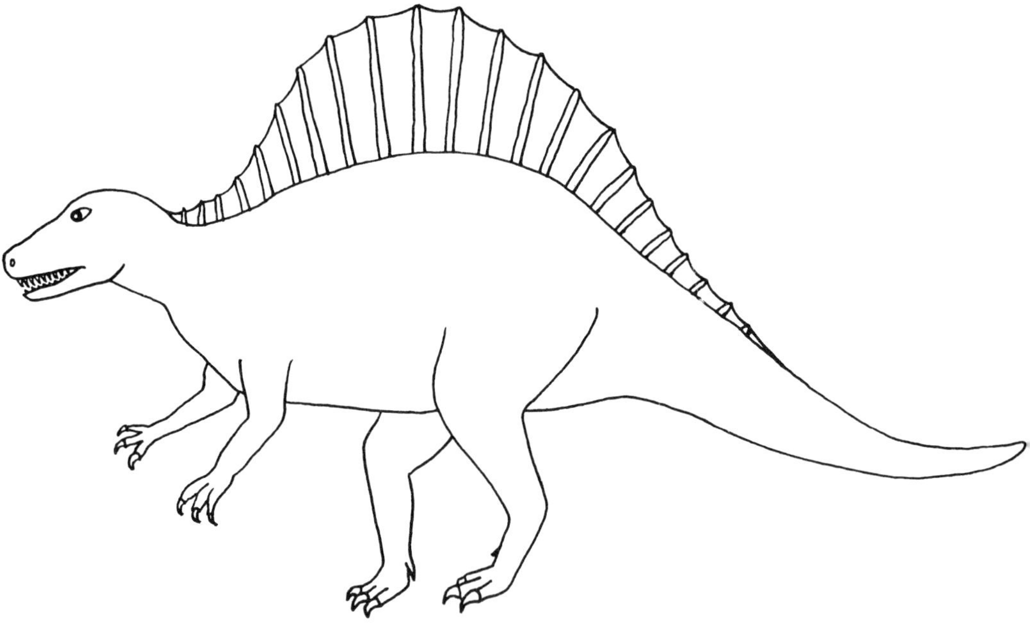

4. Which dinosaur was the 'Terrible Claw'?
5. Which is the widest known dinosaur?
6. Why would *Diplodocus* never have eaten its contemporary *Deinonychus*?
7. What was the probable purpose of the bony plates on the back of *Stegosaurus*?
8. How many toes did *Hypsilophodon* have?
9. 'All dinosaurs are extinct.' True or false?
10. Of which head-butting dinosaur is 'GO RACE TESS' an anagram?
11. Which dinosaur was typically about 10 metres long: *Compsognathus, Corythosaurus, Scelidosaurus, Heterodontosaurus* or *Brachiosaurus*?
12. Meat-eating dinosaurs like *Megalosaurus* and *Allosaurus* are grouped together in which 'infraorder' of the saurischia?
13. What is the scientific study of the *classification* of living things called?
14. In 1884 the first important dinosaur in Canada was found. Which dinosaur was it?
15. In the Ornithischia, which suborder comprises *Triceratops* and *Chasmosaurus*?
16. Correct this spelling mistake: HADD ROW SAW RUSS.
17. Was *Pachycephalosaurus* herbivorous, carnivorous or omnivorous?
18. Who is this famous dinosaur person?

19. The first *Oviraptor* was discovered in 1924 in Mongolia. Why was it named *Oviraptor philoceratops*?
20. What have the following dinosaurs got in common: *Mamenchisaurus, Lufengosaurus, Shantungosaurus, Tsintaosaurus* and *Huayangosaurus*?
21. Before the word 'fossil' came to refer to the remains of long-dead living things, it meant something much more general. What?
22. What kind of animal was *Hylonomus*?
23. Identify this dinosaur: It was 1.2 metres long. It was a South African late Triassic/early Jurassic herbivorous ornithopod, closely related to *Heterodontosaurus.* Its name means 'crane reptile'.
24. 'Dinosaurs have not yet been found in Antarctica because the freezing conditions meant that none lived there.' True or false?
25. In 1861 an important bird-like fossil was found in Solnhofen in South Germany. What was the name given to it?
26. How many toes did *Iguanodon* have?
27. What was the most striking feature of *Ouranosaurus*?
28. Which American President created the Dinosaur National Monument?
29. Why is it unrealistic to suggest that the dinosaurs died out because the mammals ate all their eggs?
30. Put this classification into the correct order by matching the taxonomic divisions against their names:

Dinosauria	Class
Reptilia	Genus
Archosauria	Suborder
Ornithopoda	Subclass
Stegosauria	Order
Stegosauridae	Superorder
Stegosaurus	Family

Quiz 10

1. In Romania, especially in the area known as Transylvania, all the dinosaurs discovered are dwarfs. What is a possible explanation for this?
2. Which family of dinosaurs was the 'head-butters'?

3. Which modern African mammals have been compared to *Hypsilophodon*?
4. Which dinosaur family was the 'duck-billed' dinosaurs?
5. Which dinosaur had the biggest head (the size of a small car) of any land animal ever known?
6. Identify this dinosaur:

7. Why would *Tyrannosaurus* never have met *Brontotherium*?
8. In 1811 a 12 year old girl, Mary Anning, discovered on the cliffs near Lyme Regis in Dorset which extinct reptile?
9. How many millions of years ago did the dinosaurs first evolve?
10. Identify this dinosaur:

11. Towards the end of which geological period might the Earth have looked something like this?

12. From which 'infraorder' in the saurischian suborder Sauropodomorpha was the presumed ancestor of the Sauropods?
13. Ornithischian dinosaurs have a 'predentary bone'. What is this?
14. '*Tyrannosaurus* was a Theropod.' True or false?
15. '*Tyrannosaurus* was a Tetrapod.' True or false?
16. Which of the following dinosaurs typically weighed 1.5 tonnes: *Stegosaurus, Triceratops, Iguanodon, Tyrannosaurus* or *Compsognathus*?
17. Which armoured dinosaur had 'many spines'?
18. Which dinosaur's name is hidden by this anagram: 'OAR TRIP CLOVE'?
19. Sauropods are often described as being totally vulnerable to attack from large carnivorous dinosaurs but this is not quite true. Apart from their large size, they did have one defensive weapon. What was it?
20. How were the plates on the back of *Stegosaurus* arranged?
21. Which dinosaur was described in 1824 as the 'Great Fossil Lizard of Stonesfield'?
22. What sort of cosmic catastrophe do some scientists believe to be responsible for killing off the dinosaurs?

23. To which of the following groups of marine reptiles did *Kronosaurus* belong: Ichthyosaurs, Plesiosaurs, Pliosaurs, Mosasaurs or Placodonts?
24. How many toes did *Triceratops* have on (a) its front legs and (b) its back legs?
25. Identify this dinosaur: It was a protoceratopsid, slightly smaller than *Protoceratops*. It was found in North America; it may have been bipedal with grasping hands and long back legs. There were no openings in its poorly developed frill. There was no horn.
26. Correct this spelling mistake: BRAKKEEO SOARUSS.
27. Was *Ornithomimus* a herbivore, a carnivore or an omnivore?
28. What does the name *Gallimimus* mean?
29. Most large dinosaurs have extra bones to reinforce their bellies. They are called 'belly ribs'. What is the scientific term for them?
30. Which Australian Iguanodontid had a bump on its nose?

Quiz 11

1. How many millions of years ago were the first fossils formed?
2. What is the most obvious difference between different species of Hadrosaurs?
3. *Cephalaspis* was an example of which kind of extinct fish?
4. Who is this famous dinosaur person?

5. Where is the impact crater left by a huge meteorite, which some researchers believe might have been the one which killed off the dinosaurs?
6. What colour are most of the rocks dating from the late Triassic?
7. Why do large ceratopids like *Triceratops* require very strong front legs?

8. What was the probable use of the crests of crested Hadrosaurs?
9. *Protoceratops* and other dinosaurs have been shown to exhibit some sexual dimorphism. What is sexual dimorphism?
10. The large *Pachycephalosaurus* had a very similar but much smaller relative. Which dinosaur was it?
11. Which dinosaur was the 'hollow form'?
12. Which is the odd-one-out: *Edmontosaurus, Parasaurolophus, Dilophosaurus, Lambeosaurus* or *Corythosaurus*?
13. Why do so many dinosaurs have openings (fenestrae) in their skulls and other bones?
14. What did *Ceratosaurus* have running down the length of its back?
15. Identify this dinosaur:

16. How many horns did *Triceratops* have?
17. Ceratopsians are divided into three families. What are they?
18. Which had the most defensive armour plating: Nodosaurs or Ankylosaurs?
19. Is *Scelidosaurus* a Triassic, a Jurassic or a Cretaceous dinosaur?
20. What are the three facial characteristics of ceratopids?
21. Which Chinese dinosaur is described as most resembling the legendary unicorn?
22. Which dinosaur suborder comprised the first herbivorous dinosaurs and the first 'high browsers'?
23. What are 'high browsers'?
24. '*Tyrannosaurus* had a brain that was bigger than the human brain.' True or false?
25. Of which pterosaur is 'ROOM HIND POD' an anagram?

26. Which is the smallest of the baby dinosaur remains so far discovered?
27. Was *Stegosaurus* a herbivore, a carnivore or an omnivore?
28. Which of the following dinosaurs weighed typically between 8 and 12 tonnes: *Stegosaurus, Triceratops, Apatosaurus, Segnosaurus* or *Tyrannosaurus*?
29. *Deinonychus* was undoubtedly a ferocious killer but, as dinosaurs go, was not particularly large; and yet it probably preyed upon dinosaurs much larger than itself. How?
30. How did *Deinonychus* probably kill animals smaller than itself?

Quiz 12

1. Which hadrosaurid dinosaur gave rise to the nickname 'duck-billed' as a result of its very flat, long, broad snout?
2. Which dinosaur family has the longest name?
3. *Protoceratops* had two very large openings in its head frill. What are the two reasons you can think of to explain this?
4. How many horns did *Styracosaurus* have?
5. Ornithomimosaurs were apparently defenceless. Although they might have been camouflaged, what was their main method of escaping their enemies?
6. *Allosaurus* had 3 fingers. How many did its relative *Ceratosaurus* have?
7. Put this classification into the correct order against the taxonomic division by matching column two against column one:

Kingdom	*Tyrannosaurus*
Phylum	Carnosauria
Class	Tyrannosauridae
Subclass	*rex*
Order	Animalia
Infraorder	Archosauria
Suborder	Chordata
Family	Theropoda
Genus	Reptilia
Species	Saurischia

8. Identify this dinosaur:

9. How many vertebrae are there in the tail of *Diplodocus*? About 50, 60, 70, 80 or 90?
10. *Massospondylus* was a prosauropod possessing gastroliths. What are gastroliths?
11. For which aspect of dinosaur studies are the following people famous: Neave Parker, Charles R. Knight, John Gurche, Zdenek Burian and Eleanor M. Kish?
12. Which is the smallest known ceratopsian?
13. Which hadrosaur had the longest crest?
14. Hypsilophodontids were originally classified as part of another dinosaur family until they were given a family of their own. Of which family were they once a part?
15. What was the probable purpose of the frills of ceratopids?
16. What advantage could the possession of a sail have for the mammal-like reptile *Dimetrodon*?
17. What was *Archelon*?

18. Identify this pterosaur:

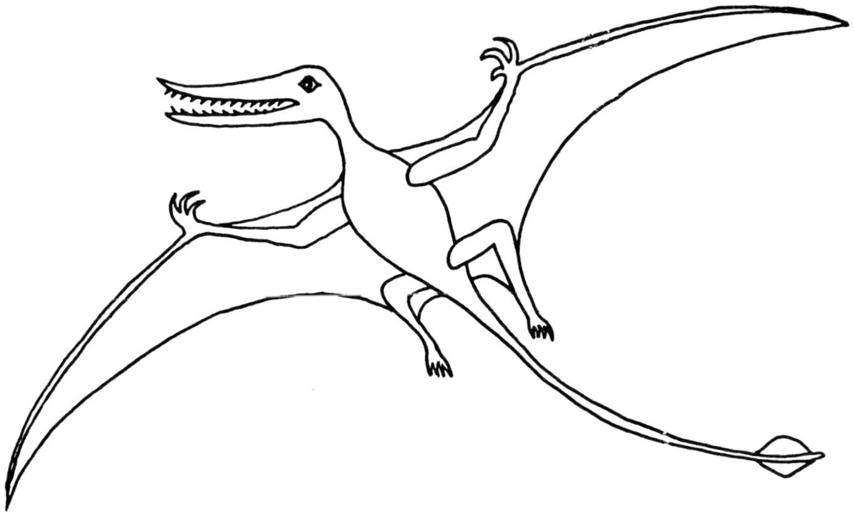

19. Sometimes a dinosaur can acquire many different names. For example the nodosaur *Struthiosaurus* has been given over 10 different names, one for almost every time it has been found. Why should this be so?
20. Of which prosauropod dinosaur is 'LOUT SUES PARA' an anagram?
21. To which ornithischian suborder does *Saurolophus* belong?
22. Which is the odd-one-out: *Heterodontosaurus, Psittacosaurus, Centrosaurus, Edaphosaurus* or *Corythosaurus*?
23. Some older books on dinosaurs mention a dinosaur named *Laelaps*, discovered in 1866. Most more recent books decline to mention it. What has happened to *Laelaps*?
24. Correct the following spelling mistake: 'DINE ON YOU, CHUCK!'
25. What is the name of the lobe-finned fish believed to be extinct for 65 million years that was rediscovered, alive, off the coast of Madagascar in 1938?
26. One theory is that a large meteorite striking the Earth was responsible for finishing the dinosaurs; another that comets, which strike the Earth regularly every 26 million years, were the cause. Regarding this possibility, what is the Oort cloud?

27. Why do large ceratopian dinosaurs have shorter tails than other dinosaurs of a similar size?
28. Which dinosaur had elongated chevron bones in its tail?
29. What is the controversy regarding *Monoclonius* and *Centrosaurus*?
30. What is the difference between a ceratopid, a ceratopsid, a ceratopian and a ceratopsian?

Quiz 13

1. Contrary to the information supplied by some dinosaur books, why is *Pentaceratops* a misnomer?
2. When *Stegoceras* was about to head-butt another *Stegoceras* what position would it have assumed for its backbone?
3. What does the name *Compsognathus* mean?
4. Was *Compsognathus* a herbivore, a carnivore or an omnivore?
5. Which of the following dinosaurs typically weighed about three tonnes: *Stegosaurus, Stegoceras, Anatosaurus, Apatosaurus* or *Anchisaurus*?
6. This is a diagram of a dinosaur bone. Which one?

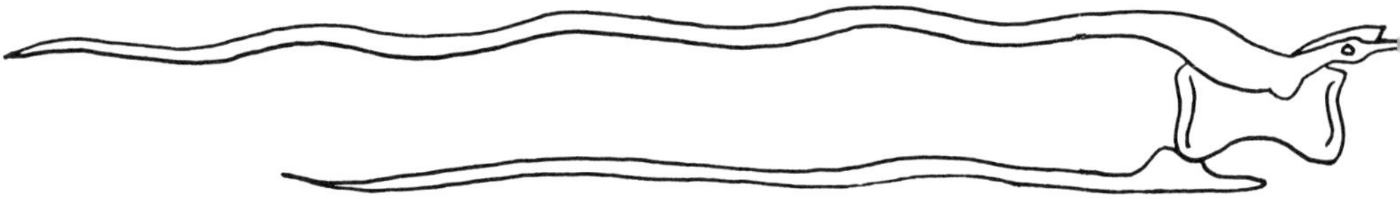

7. What was the probable purpose of the spines of *Spinosaurus*?
8. Which family of dinosaurs had the most teeth?
9. Of which ornithopod dinosaur is 'TO SHOOT A RUDE NURSE' an anagram?
10. To which saurischian infraorder does *Plateosaurus* belong?
11. Which were the first two dinosaurs to be found in America, in 1855 in Montana, and who named them?
12. Many specimens of *Coelophysis* have been found with small *Coelophysis* bones within them. At first it was believed that they were unborn young inside a pregnant female but what is now regarded as the most likely explanation?

13. Identify this dinosaur:

14. What is the name given to the small bones which surround the outer edge of the frill of *Triceratops* and which gave it its characteristic wavy rim?
15. Most pachycephalosaurids are known solely from the roof of the skull. Why should only this part survive?
16. What was the defence of *Stegosaurus*?
17. Identify this famous dinosaur person:

18. In order to avoid crushing her own eggs, how is it believed that a large sauropod laid them?
19. Which was the largest member of the Hypsilophodontidae?
20. What are the two main differences between the skulls of the male and female *Protoceratops*?
21. Who wrote the famous novel 'The Lost World' about a plateau in South America populated by dinosaurs?
22. What were the two different defensive strategies displayed by a nodosaur and an ankylosaur against a large predator like *Tyrannosaurus*?
23. What was unusual about the tail bones of *Ichthyosaurus*?
24. '*Pteranodon* was the largest pterosaur.' True or false?
25. In the Jurassic period, was Africa part of Gondwanaland or Laurasia?
26. Tyrannosaurids have two fingers. Which is the only non-Tyrannosaurid dinosaur to share this characteristic?
27. Identify this marine reptile:

28. Correct this spelling mistake: ORNIFF OLLIES TEASE.
29. Which was the largest land mammal that ever lived?
30. What is 'Dinoturbation'?

Quiz 14

1. Which was the 'west wind' dinosaur?
2. The frill of *Protoceratops* provided an attachment area for 'capiti-mandibularis'. What was this?

3. Although hadrosaurs *could* be divided into three groups (non-crested, solid-crested and hollow-crested), what are the two subfamilies into which they are normally placed?
4. It is believed by some that *Lesothosaurus* aestivated. What is 'to aestivate'?
5. What is the name of the strange bone that is in the eye cavity of most ornithischian dinosaurs?
6. Identify this dinosaur:

7. Which suborder in Ornithischia or infraorder in Saurischia possesses the greatest number of genera?
8. Which dinosaurs were the 'slow reptiles'?
9. Which is the odd-one-out: *Diplodocus, Apatosaurus, Mamenchisaurus, Plateosaurus* or *Cetiosaurus*?
10. What kind of animal was *Elasmosaurus* and what was its connection with the famous rivalry between Cope and Marsh?
11. What is distinctive about the frill of *Styracosaurus*?
12. *Psittacosaurus* was closely related to the ceratopsids but lacked which important ceratopsid characteristic?
13. Hadrosaurs had no means of defence and could not outrun predators so how is it believed that they escaped from their enemies?
14. To which Jurassic land mass did Europe belong?
15. *Leptoceratops* was a very similar size to *Protoceratops*. What does its name mean?

16. Identify this hadrosaur:

17. Of which carnosaur is 'MA'S A RUG LOUSE' an anagram?
18. To which ornithischian suborder does the dinosaur *Torosaurus* belong?
19. To which saurischian infraorder does the dinosaur *Syntarsus* belong?
20. Which of the following dinosaurs was 30 metres long: *Stegosaurus, Cetiosaurus, Supersaurus, Diplodocus* or *Apatosaurus*?
21. Identify this famous dinosaur person:

22. *Deinocheirus* was possibly a huge ornithomimosaur but it is known only from which part of its anatomy?
23. In quarries which have yielded large numbers of *Deinonychus* there have also been the remains of a large herbivore. What was this presumed prey species for *Deinonychus*?

24. What do the bones of the huge sauropods have in common with the bones of little birds, and what was the identical benefit derived?
25. Which of the following is a specific name for *Megalosaurus*: *rex, carnegii, bucklandi, walkeri* or *willistoni?*
26. Why did Pachycephalosaurs head-butt one another?
27. *Styracosaurus* is regrettably known only from its skull. What part of the skull produces the marvellous spikes on its frill?
28. When the dinosaurs became extinct, they were accompanied into extinction by many other major groups of animals. Name four of these groups.
29. What was the advantage to a large herbivorous dinosaur of having a long neck?
30. How would *Tyrannosaurus* probably have dealt with an Ankylosaur or a Nodosaur given that both these two dinosaurs were heavily armoured?

Quiz 15

1. Why was *Camptosaurus* called the 'bent or flexible reptile'?
2. Was *Camptosaurus* a herbivore, a carnivore or an omnivore?
3. Which of the famous expeditions in the past was led by Edwin Henning and Werner Janensch?
4. Fast running (fleet-footed) dinosaurs all have very long tails which they hold out straight and rigid. What advantage would such a posture have?
5. Which of the following dinosaurs weighed typically 30 tonnes: *Apatosaurus, Tyrannosaurus, Iguanodon, Brachiosaurus* or *Cetiosaurus?*
6. Of which ceratopsian dinosaur is 'A TENT CAP SPORE' an anagram?
7. To which ornithischian suborder does *Ouranosaurus* belong?
8. Which dinosaur had para-sacral spines?
9. '*Iguanodon* has been reported from nearly all over the world, but only remains in Europe and Asia have been substantiated.' True or false?
10. Which small dinosaur was named 'hollow tail' because its tail vertebrae were hollow?
11. What was unusual about the hand of *Ornitholestes*?

12. Identify this dinosaur:

13. Which dinosaur made these footprints: *Apatosaurus, Iguanodon, Tyrannosaurus, Diplodocus* or *Stegosaurus?*

14. How did *Deinonychus* keep its tail rigid when running?
15. How many toes did *Plateosaurus* have on (a) its front legs and (b) its back legs?

16. What is happening in this picture and when did it take place?

17. Why did some species of dinosaur survive in India during the Cretaceous when they had become extinct elsewhere?
18. Many dinosaurs had a ring of bones surrounding and supporting the eyeball. What is this ring called?
19. Of which ornithopod dinosaur is 'HID POISON HYPO' an anagram?
20. In 1970 the Canadian theropods were reviewed by Dale Russell. What happened to *Gorgosaurus*, the dragon reptile?
21. How many fingers, including the thumb spike, had *Iguanodon*?
22. Hypsilophodontids were closely related to fabrosaurids but they differed in the shape of their teeth and jaws and in one other respect, which prevented them from wasting food while chewing. What was this?
23. Why was *Pterodactylus* so named?
24. Did Ichthyosaurs lay eggs, like dinosaurs?
25. Which, at 2 metres long, was the smallest nodosaurid?
26. Which is the earliest-known stegosaurid?
27. What is unusual about the frill of *Centrosaurus*?

28. Who proposed that the crests of hadrosaurs were signalling devices?
29. What does the name *Maiasaura* mean?
30. What is remarkable about *Maiasaura*?

Quiz 16

1. Who is this famous dinosaur person?

2. *Struthiomimus* had long toes but its metatarsals, the bones of the upper foot, were very long indeed, almost as long as the lower leg bones. What does this tell us about the *Struthiomimus* lifestyle?
3. Which dinosaur had the longest and biggest teeth known?
4. In America, there is the 'Dinosaur National Monument', but in the same state there is another, similar site, protected as a 'natural landmark' since 1967. What is the name of this place?
5. On which toe is the claw of *Deinonychus*?
6. One diplodocid was virtually identical to *Diplodocus* except that its neck vertebrae were longer. Which dinosaur was this?
7. *Plateosaurus* remains are the most abundant of the early dinosaurs. Where in Germany have large quantities been found?

8. Identify this pterosaur:

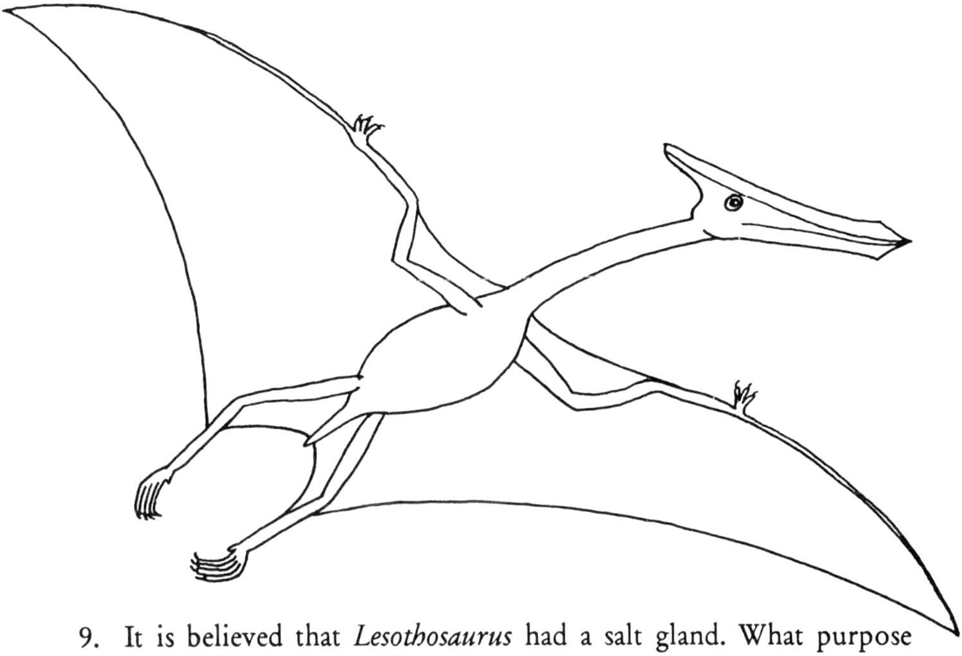

9. It is believed that *Lesothosaurus* had a salt gland. What purpose would this have served?
10. Which iguanodontid had the longest head?
11. To which ornithischian suborder does *Lexovisaurus* belong?
12. Of which hadrosaur is 'OH, A SOUR CRUSTY' an anagram?
13. Of which pterosaur is 'ORNATE POND' an anagram?
14. To which ornithopod family does *Lambeosaurus* belong?
15. Palaeontologists used to believe that sauropods had an aquatic existence. The problem of their being able to inflate their lungs against the water pressure has been mentioned in an earlier quiz but why would the huge pillar-like legs of sauropods also suggest that they lived on land?
16. How long was a typical *Hypsilophodon*: 1, 2, 3, 4 or 5 metres long?
17. How many horns did *Pentaceratops* have?
18. Ceratopids are divided into long- and short-frilled categories. Why could *Triceratops* belong in both categories?
19. Which dinosaur family shared a behavioural characteristic with sheep and goats?
20. What is the connection between the element iridium and the death of the dinosaurs?

21. Identify this dinosaur:

22. Which stegosaurid appears to have been the last surviving member of its family and where was it found?
23. Which relative of *Scelidosaurus* was found in Portugal?
24. *Nodosaurus* had plates of bone covering the head. This is described as 'dermal bone'. What is 'dermal bone'?
25. Which pterosaur is regarded as being a plankton feeder because of its comb-like teeth?
26. 'The mass extinction of many life forms at the end of the Cretaceous is the only time such an event has taken place.' True or false?

27. Correct this spelling mistake: PROW TOE SEER A TOSP.
28. What kind of animal was *Deinosuchus*?
29. After the Mesozoic came the Cenozoic era. What does 'Cenozoic' mean?
30. Identify this hadrosaur.

Quiz 17

1. Which ankylosaurid had the knobbliest head?
2. Which dinosaur family lived on a diet that included pine needles?
3. What do the following dinosaur names have in common: *Fabrosaurus, Lambeosaurus, Parksosaurus, Othnielia* and *Gilmoreosaurus*?
4. A section cut across one of the plates of *Stegosaurus* does not reveal solid bone. Why, instead, do we find a honeycomb of spaces which clearly had a good blood supply?
5. Identify this dinosaur:

6. What do we know about the colour of dinosaurs' skins?
7. What was unusual about the hands of *Compsognathus*?
8. What are the two main groups into which the pachycephalosaurians can be divided?
9. What is palaeomagnetism and how is it useful in the study of dinosaurs?
10. Which of the following dinosaurs was typically 1.8 metres long: *Compsognathus, Stegoceras, Protoceratops, Ornithomimus* or *Diplodocus*?
11. In 1858 the first hadrosaur was found in America. It was a partial skeleton but its limbs were intact. Which fact was immediately realised about the dinosaur that had previously eluded the English in over thirty years of work?
12. To which ornithischian suborder does *Hypsilophodon* belong?
13. '*Stegosaurus* was extinct in North America throughout the Cretaceous.' True or false?
14. Which one of the following statements is true?
    (a) Dinosaurs were synapsid reptiles.
    (b) Dinosaurs were the only diapsid reptiles.
    (c) Dinosaurs and lepidosaurs were diapsid reptiles.
    (d) Dinosaurs and ichthyosaurs were euryapsid reptiles.
15. Which of the following dinosaurs weighed typically between four and five tonnes: *Stegosaurus, Triceratops, Iguanodon, Tyrannosaurus* or *Apatosaurus*?
16. Why was *Ceratosaurus* named the 'horn reptile'?
17. Which iguanodontid had a sail similar to that of *Spinosaurus* or *Dimetrodon*?
18. Who discovered *Supersaurus* and *Ultrasaurus*?
19. What is the unique and incredible feature about both *Supersaurus* and *Ultrasaurus*?
20. Which botanical event coincided with the evolutionary success of the hadrosaurs?
21. Which dinosaur most resembles a modern rhinoceros?
22. Which is the odd-one-out: *Hypsilophodon, Brachiosaurus, Compsognathus, Heterodontosaurus* or *Saltopus*?
23. Iguanodontids only had teeth at the back of their jaws. What was at the front?
24. Where were the famous fossils of *Protoceratops* found?

25. How many toes did *Stegosaurus* have on (a) its front legs and (b) its back legs?
26. Was *Stegoceras* a high-domed or a low-domed pachycephalosaur?
27. Which of the following dinosaurs made these footprints: *Brachiosaurus, Iguanodon, Triceratops, Diplodocus* or *Ankylosaurus*?

28. What were the two main differences between a plesiosaur and a pliosaur?
29. Which geological period lasted from 190/200 to 135/136 million years ago?
30. Of which close relative of the ceratopsidae is 'STRATA COUP IS US' an anagram?

Quiz 18

1. Identify this dinosaur:

2. There is much discussion about what kind of heart a dinosaur had. An amphibian like a frog has an 'undivided heart', while a mammal or bird has a 'fully-divided heart'. What kind of heart do modern reptiles have?
3. What kind of heart is it almost certain that dinosaurs had?
4. What does the name *Pteranodon* mean?
5. What was the typical length of the head of *Kronosaurus*, a marine pliosaur: 1.9m, 2.4m, 2.9m, 3.5m or 4m long?
6. Which group of dinosaurs had a rostral bone at the tip of the upper jaw?
7. We now know that the sail of *Dimetrodon* was used as a means of controlling its temperature but how was it originally believed to help with its supposed aquatic existence?
8. Which is the earliest-known nodosaurid from the early Jurassic?
9. *Stegosaurus* probably weighed about 1.5 tonnes but it had a ridiculously small brain. How much did its brain weigh?
10. Identify this hadrosaur:

11. To which saurischian infraorder does *Massospondylus* belong?
12. A recent find in 1981 in Australia was at first reported to be *Iguanodon* until it was realised how very different it was, namely that the creature had a bump on the nose. What was the name given to it?
13. When and where were the first dinosaurs found in Africa?
14. *Petrolacosaurus* was not a dinosaur but why is it relevant to dinosaur history?

15. 'The plates of *Stegosaurus* were attached to its vertebrae.' True or false?
16. In 1972 a new species of *Compsognathus* was described. *Compsognathus corallestris* was larger than the other species but how else was it supposed to differ?
17. Of which Jurassic land mass was Antarctica a part?
18. What is the new name, proposed in 1989, for *Centrosaurus*?
19. Which hadrosaurs could be described as 'saurolophine'?
20. What was missing from the head of *Pachyrhinosaurus*?
21. Which of the following statements is not true:
    (a) *Gasosaurus* almost certainly preyed upon *Shunosaurus*;
    (b) *Deinonychus* almost certainly preyed upon *Tenontosaurus*;
    (c) *Velociraptor* almost certainly preyed upon *Protoceratops*;
    (d) *Sarcolestes* almost certainly preyed upon *Hadrosaurus*?
22. Identify this dinosaur contemporary:

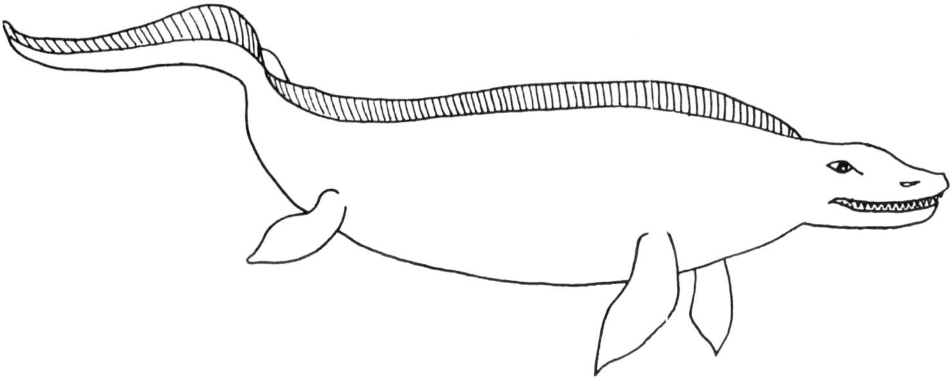

23. Of which small dinosaur is 'SO POLICE SHY' an anagram?
24. To which ornithischian suborder does *Nodosaurus* belong?
25. Which of the following dinosaurs typically weighed 78 tonnes: *Brachiosaurus, Ultrasaurus, Triceratops, Apatosaurus* or *Diplodocus*?
26. *Oviraptor* had a bone possessed by very few other dinosaurs. It is characteristic of birds. What is it?
27. Correct this spelling mistake: PARASOLLE OFF US.
28. Which Mongolian Ankylosaurid was the 'Basket Tail'?
29. Dinosaurs like hadrosaurs and ceratopids possessed cheeks. What advantage did this give them over cheekless dinosaurs like *Tyrannosaurus*?
30. Which ceratopid had no epoccipitals ?

Quiz 19

1. Why is it doubtful that the ceratopian frill was a defensive shield?
2. Who originated the theory that *Lesothosaurus* and *Heterodontosaurus* aestivated?
3. *Bothriospondylus* is a sauropod found in Europe and Madagascar whose remains amount to little more than pieces of backbone and leg. Why is *Bothriospondylus* a particularly appropriate name?
4. Iguanodontids were widespread during the late Jurassic and early Cretaceous but towards the end of the Mesozoic were confined to which continent?
5. Identify this dinosaur:

6. What was the probable behavioural difference between the low-domed and the high-domed pachycephalosaurians?
7. What have these dinosaurs in common: *Lesothosaurus*, *Albertosaurus*, *Dravidosaurus*, *Nemegtosaurus* and *Itemirus*?

8. Here are three stegosaurid plates belonging to *Stegosaurus*, *Kentrosaurus* and *Tuojiangosaurus*. Which is which?

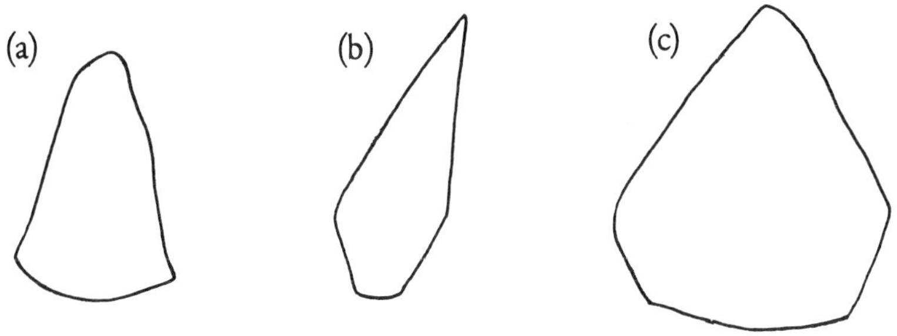

9. Which is the closest-known relative to *Tyrannosaurus*?
10. 'Hypsilophodontids are a family of dinosaurs found all over the world.' True or false?
11. Of which close relative of the armoured dinosaurs is 'AL ROUSE DISCUS' an anagram?
12. To which ornithischian suborder does *Fabrosaurus* belong?
13. What was the most striking feature of *Dilophosaurus*?
14. Identify this famous dinosaur person:

15. What is the name of the recently discovered (1965) sauropod from Mongolia which had a massive pelvis and a unique tail consisting of vertebrae fitting together with immensely strong ball-and-socket joints?

16. What are the main differences between the hand of a hadrosaurid and that of an iguanodontid?
17. Which hadrosaur group had hollow crests?
18. In which continent have remains of *Leptoceratops* been found?
19. Which major dinosaur group flourished as the stegosaurs declined and yet remained absent from India where stegosaurs continued to the end of the Cretaceous?
20. What is the evidence to support the idea that ankylosaurids had a keen sense of smell?
21. Which pterosaur had a long bony crest on the back of its head?
22. Identify this dinosaur contemporary:

23. All known Pachycephalosaurians are from the Northern Hemisphere with one exception. This is *Majungatholus*. In which country was it found?
24. *Centrosaurus* and *Monoclonius* shared an anatomical feature with *Iguanodon* and hadrosaurs. This was the coronoid process at the back of the lower jaw. What was it for?
25. Along the back of *Stegosaurus* where were (a) the smallest plate and (b) the largest?
26. Which Polish doctor is responsible for describing many of the more recent finds from Mongolia: Dr Magdalena Borusk-Bialynicka, Dr Krystof Penderecki, Dr Wojciech Jaruzelski, Dr Marian Miskiewicz or Dr Henryk Jablonski?

27. What is Ichnology, more correctly termed Palaeoichnology?
28. Towards the end of the Permian, the pelycosaurs died out and were replaced by which other group of mammal-like reptiles?
29. Which American wrote a book in 1858 about dinosaur footprints which he believed were made by giant birds?
30. Which bone, at the back of the skull in reptiles, is greatly enlarged in ceratopians to form the frill?

Quiz 20

1. Identify this dinosaur:

2. Which geological period lasted from 135 to 65 million years ago?
3. What were the Aëtosaurs?
4. With regard to a dinosaur like *Brachiosaurus* why is it virtually certain that dinosaurs had a fully-divided heart?
5. How is the possession of a fully-divided heart an argument in favour of warm-bloodedness?
6. Attractive though the idea of cosmic catastrophe is, what reason do most palaeontologists still favour to explain the demise of the dinosaurs?

7. Anyone interested in fossils will have seen ammonites, but what kind of animals were they?
8. What is an obvious difference between the frill of *Triceratops* and that of *Chasmosaurus*?
9. Which controversy was finally resolved by the discovery of the fossil pterosaur *Sordes pilosus* in Kazakhstan in 1970?
10. Sauropod dinosaurs had feet roughly resembling those of which present-day mammal?
11. In what way does the Chinese prosauropod *Yunnanosaurus* differ from all other known prosauropods?
12. What is the African equivalent of America's Dinosaur National Monument?
13. Which is the latest-known diplodocid, which survived to the end of the Cretaceous in Mongolia?
14. Which dinosaur family seems to have competed with, and eventually replaced, the iguanodontids world-wide (except for Europe)?
15. Identify this ceratopid:

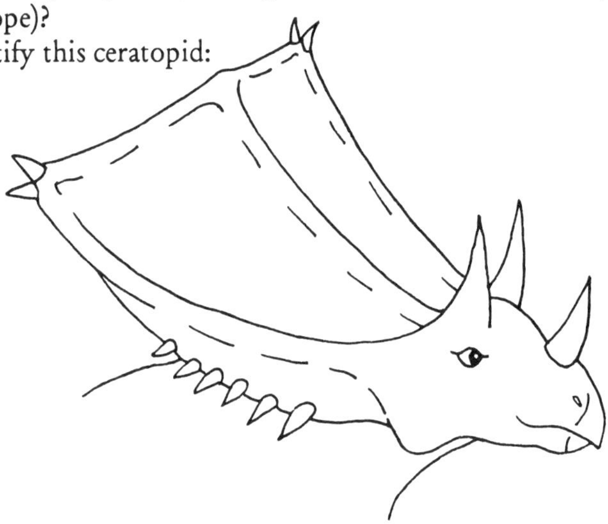

16. Which family of dinosaurs is regarded as the most 'advanced' of the ornithopods?
17. Was *Torosaurus* a long- or short-frilled ceratopid?
18. Which of the following dinosaurs was typically 9 metres long: *Stegosaurus*, *Parasaurolophus*, *Diplodocus*, *Ornithomimus* or *Protoceratops*?
19. To which saurischian infraorder does *Titanosaurus* belong?
20. To which family does *Gallimimus* belong?

21. Of which ornithischian dinosaur is 'ARCH CAUSE HAPPY SOUL' an anagram?
22. Identify this carnosaur:

23. Why was Edward Cope wrong to name three separate species of Coelophysis?
24. Which of the following Chinese dinosaurs is an important early Sauropod: *Gasosaurus, Chinshakiangosaurus, Tuojiangosaurus, Tsintaosaurus or Tatisaurus*?
25. What are zygapophyses?
26. *Avimimus* was described by the Russian Dr Kurzanov in 1981. What is special and controversial about it?
27. *Baryonyx* had a claw that was three times longer than that of *Deinonychus*, but what was the difference between the position of their claws?
28. Which group of dinosaurs can be divided according to whether their teeth are 'spatulate' or 'cylindrical'?
29. How many vertebrae are there in the tail of *Apatosaurus*?
30. Why don't most dinosaur books mention *Diceratops*?

Quiz 21

1. Who wrote the 1990 bestseller 'Jurassic Park', made into a Steven Spielberg film about a dinosaur theme-park?
2. Why might *Opisthocoelicaudia* one day become *Nemegtosaurus*?
3. Which is the earliest-known pachycephalosaur and where was it found?

4. Match these four dinosaurs with the meanings of their names:
   (a) *Lanasaurus* (1) marsh reptile (a Hadrosaurid)
   (b) *Telmatosaurus* (2) unmarried reptile (a Tyrannosaurid)
   (c) *Alectrosaurus* (3) marvellous (a ceratopsid)
   (d) *Agathaumas* (4) wool reptile (a Heterodontosaurid)
5. Identify this dinosaur:

6. Most dinosaurs have their heads positioned on their necks so that the most natural position for the head would be one of looking at the world. In this respect, what was different about the natural position of the head of *Stegoceras*?
7. How many horns did *Anchiceratops* possess?
8. What was the name of the very early mammal found in Wales?
9. What is the eventual fate of all species of living things?
10. Which of the following dinosaurs possessed this tooth: *Tyrannosaurus, Pentaceratops, Stegosaurus, Diplodocus* or *Deinonychus*?

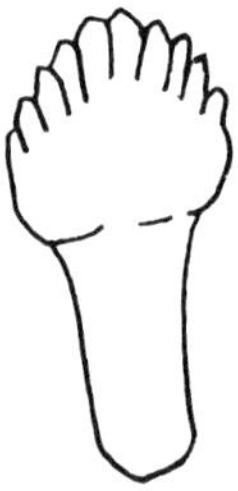

11. Explain how the sail of *Spinosaurus* could be used as (a) a solar panel or (b) a radiator?
12. What is the alternative name for the Cenozoic era?
13. *Troödon* used to be thought of as a pachycephalosaur and then as similar to *Hypsilophodon* but in a family of its own. It is now believed to be one of the most intelligent dinosaurs. Which family has been renamed the Troödontidae to accomodate it?
14. If *Troödon* had remained in the Ornithischia, what would have been so unusual about it?
15. In the film 'Jurassic Park' which two characteristics (one anatomical, one behavioural) of *Dilophosaurus* are completely fanciful?
16. The following four words are 'dinosaur' in French, German, Spanish and Italian. Which is which: (a) Dinosaure; (b) Dinosaurio; (c) Dinosauro; (d) Dinosaurier?
17. Identify this hadrosaur:

18. To which ornithischian suborder does *Acanthopholis* belong?
19. Of which stegosaurid is 'ASSURE NO TURK' an anagram?
20. What is the 'dinosauroid' invented by Dale Russell?
21. The American Barnum Brown collected fossils in the Red Deer River Valley, Canada in 1910. Which Canadian family copied his technique of sailing down river in a huge barge?
22. Who discovered *Compsognathus* and where?
23. 'Protoceratopids have only been found in Asia.' True or false?
24. Name two of the four basic characteristics possessed by coelurosaurs which set them apart from other dinosaur groups.

25. Which of the following dinosaurs made these footprints: *Diplodocus, Tyrannosaurus, Iguanodon, Stegosaurus* or *Compsognathus*?

26. Who first described *Pentaceratops* and when?
27. Were dinosaurs, as far as we know, 'oviparous', 'ovoviviparous' or 'viviparous'?
28. To which saurischian infraorder does *Camarasaurus* belong?
29. Which of the following dinosaurs weighed typically about 6 tonnes: *Brachiosaurus, Tyrannosaurus, Stegosaurus, Triceratops* or *Hypsilophodon*?
30. Name the important Chinese quarry at Zigong, in Sichyan Province, which has yielded so many interesting new dinosaurs since 1979.

Quiz 22

1. Which one of the following groups of plants could never have been eaten by dinosaurs and why: Cycads, Horsetails, Ginkgos, Ferns, Conifers or Grasses?
2. The jaws of *Dilophosaurus* were much weaker than those of *Ceratosaurus*, yet these carnivorous dinosaurs were the same size. What does this tell us about their feeding habits?
3. How many vertebrae are in the neck of *Diplodocus*: 11, 12, 13, 14 or 15?
4. What is special about the first finger on the hand of *Plateosaurus*?
5. Why is the name *Arrhinoceratops* a misnomer?
6. What have the following dinosaurs got in common: *Spinosaurus, Altispinax, Acrocanthosaurus, Ouranosaurus* and *Metriacanthosaurus*?

7. Which famous dinosaur person is this?

8. Identify this dinosaur:

9. Of which Chinese dinosaur is 'TRUSTS AS AN IOU' an anagram?
10. What is taphonomy?
11. Which of the following dinosaurs was typically 27 metres long: *Diplodocus, Supersaurus, Ornithomimus, Brachiosaurus* or *Tyrannosaurus*?
12. The skull of *Torosaurus* was examined in the 1930s by Dr Roy L. Moodie, a bone specialist. Which disease did he conclude that the animal suffered from?
13. To which saurischian infraorder does *Segnosaurus* belong?
14. To which ornithischian suborder does *Dravidosaurus* belong?

15. Identify this dinosaur:

16. What is unusual about the hip bones of *Ceratosaurus*?
17. Large dinosaurs reduce their weight by having holes or 'windows' in their bones. What are these holes called when they are in the vertebrae?
18. Which dinosaur was originally named by Marsh as a recently-extinct bison, *Bison alticornis*?
19. Which of the following dinosaurs possessed this tooth: *Cetiosaurus, Stegosaurus, Tyrannosaurus, Ceratosaurus* or *Struthiomimus*?

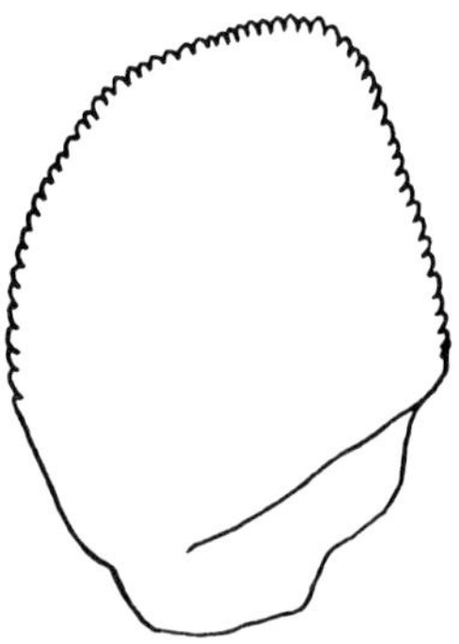

20. The first stegosaurid to be found was named *Omosaurus* by Richard Owen in 1875. Why do very few books on dinosaurs mention it these days?
21. What is distinctive about the frill of *Anchiceratops*?
22. What was the name of the tiny bipedal protoceratopid from Mongolia?
23. Who drew Gideon Mantell's attention to the fact that the fossil teeth he was studying closely resembled those of an iguana?
24. What is the connection between *Corythosaurus casuaris* and *Corythosaurus intermedius*?
25. Which is the latest-known hypsilophodontid, placed in a family of its own, which survived to the end of the Cretaceous?
26. Which group of hadrosaurs had no crests?
27. Which was the protoceratopid from Montana?
28. What was *Protiguanodon?*
29. To which family was *Scutellosaurus* formerly believed to belong?
30. Twelve genera of the Tyrannosauridae have been named on the classification chart. Which four are considered to be definite (i.e. not dubious)?

Quiz 23

1. Identify this dinosaur:

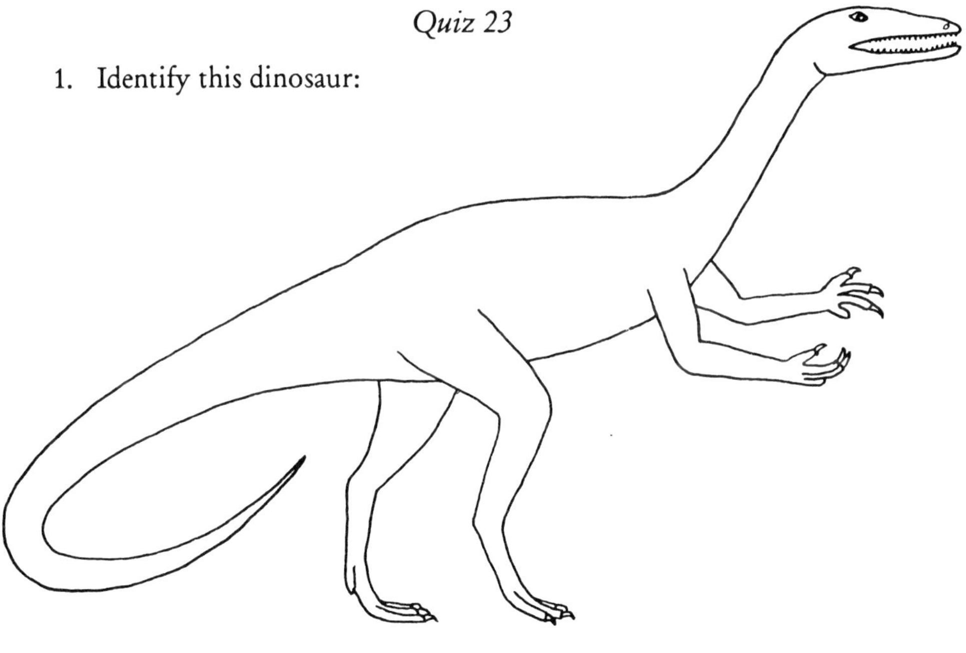

2. In which Mesozoic period did *Tyrannosaurus* live?
3. What was a probable difference between male and female *Heterodontosaurus*?
4. Name the tiny prosauropod hatchling that was discovered in Southern Argentina, whose name means 'mouse reptile'.
5. After the dinosaurs, there evolved many large species of mammals. What was the name given to the giant ground sloth?
6. Correct this spelling mistake: GALLEY MEEM UZ.
7. Identify this ceratopid:

8. To which ornithischian suborder does *Stegoceras* belong?
9. Who wrote and illustrated the delightful 1992 book 'Dinotopia'?
10. To which ornithopod family does *Zephyrosaurus* belong?
11. Which three dinosaurs were collected in Africa between 1908 and 1912?
12. Large mammal herbivores, such as cows, can move their lower jaws sideways against a fixed upper jaw and so chew plant fibres efficiently. Large ornithopod dinosaurs like *Iguanodon*, however, can only move their lower jaw up or down, so how did they manage to chew plant fibres efficiently?
13. What kind of animal was *Arsinotherium*?
14. What does the adjective 'antediluvian' mean?

15. Identify this hadrosaur:

16. Dinosaurs were probably polyphyletic. Does this mean: (a) they evolved from two or more sources; (b) they collected stamps; (c) they suffered from polyphyteltism, a rare bone disease; or (d) they came from Polynesia?
17. 'Sauropods were graviportal.' True or false?
18. 'Ornithomimosaurs were not cursorial.' True or false?
19. Which of the following dinosaurs possessed this tooth: *Stegosaurus, Iguanodon, Triceratops, Gallimimus* or *Tyrannosaurus*?

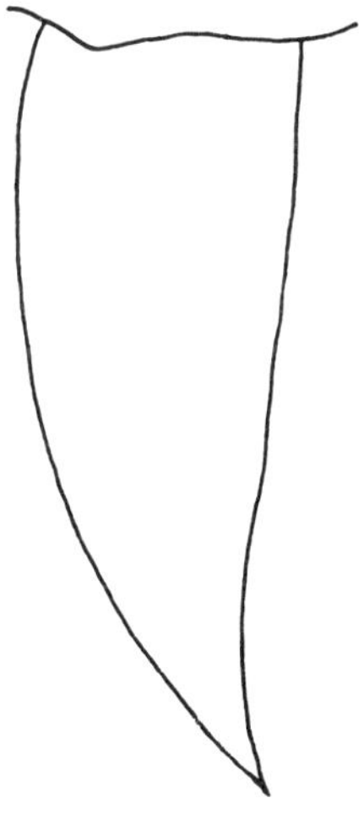

20. How many new dinosaurs were discovered or named by the great rivals Cope and Marsh? Was it 106, 116, 126, 136 or 146?
21. Which was the largest-known early fish with jaws?
22. The name *Sarcolestes* has been given to a nodosaurid. What is inaccurate about the name?
23. *Scelidosaurus* seems to be midway between which two ornithischian suborders, although it is sometimes classified as belonging to one of them?
24. Was *Pachyrhinosaurus* a short-frilled or long-frilled ceratopian?
25. Of which Jurassic land mass was India a part?
26. When and where was the first nest of *Maiasaura* discovered?
27. Which hypsilophodontid was named after O. C. Marsh?
28. How do hadrosaur teeth suggest that hadrosaurs did not live on a diet of soft vegetation?
29. Which of the following dinosaur genera has a species named *carnegii*: *Tyrannosaurus, Hypsilophodon, Diplodocus, Megalosaurus* or *Iguanodon*?
30. Who wrote the famous book 'Dragons of the Air' published in 1901?

Quiz 24

1. Identify this dinosaur:

2. Which is the odd-one-out: *Muttaburrasaurus, Saltasaurus, Segisaurus, Tarbosaurus* or *Itemirus*?
3. Of which ostrich dinosaur is 'MUSHROOM IN IT' an anagram?
4. To which ornithischian suborder does *Maiasaura* belong?
5. Who was O. C. Marsh's rich uncle who gave a large endowment to Yale University on the condition that they made him Professor of Palaeontology?
6. What are 'trace fossils'?
7. Which is the largest-known hadrosaur, on display in the Natural History Museum in Peking?
8. '*Brachiosaurus* was plantigrade.' True or false?
9. '*Hypsilophodon* was digitigrade.' True or false?
10. Which sauropod was found in Zimbabwe (formerly Rhodesia) in 1972?
11. Identify this ceratopid:

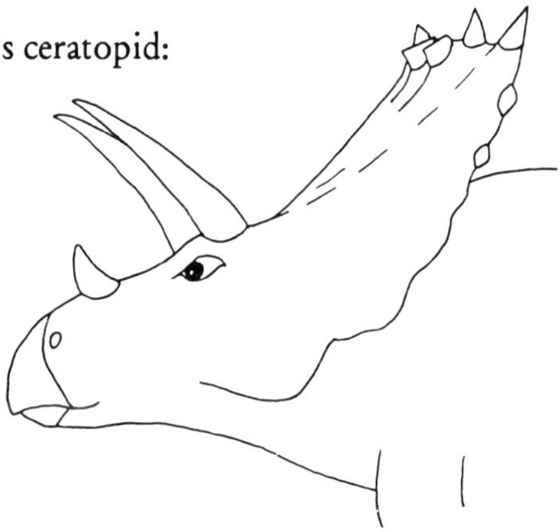

12. In 1992 which dinosaur was impounded by the American F.B.I.?
13. In what way did the backbone of *Cetiosaurus* differ from that of other sauropods?
14. To which saurischian infraorder does *Carcharodontosaurus* belong?
15. How many specimens of *Iguanodon* were found in the Bernissart coal mine, in 1878?
16. Which of the following dinosaurs typically weighed 6 tonnes: *Apatosaurus, Triceratops, Stegosaurus, Compsognathus* or *Brachiosaurus*?
17. Which is perhaps the oldest-known ornithomimosaur from Africa, discovered in 1920 and whose name means 'lightweight reptile'?

18. What do these dinosaurs have in common: *Velociraptor, Dromaeosaurus, Deinonychus, Saurornithoides* and *Stenonychosaurus*?
19. What was the typical length of the neck of *Mamenchisaurus*? Was it 6, 7, 8, 9 or 10 metres long?
20. What are the three principal features which distinguish a large carnosaur from smaller meat-eaters such as *Velociraptor*?
21. What is the relationship between *Brachiosaurus* and the appropriately-named dinosaur *Giraffatitan*?
22. Which Indian sauropod had 'a big leg'?
23. What is unusual about the lower leg bones of *Heterodontosaurus*?
24. All large ornithopods had extra bones in their vertebrae arranged like a trellis, criss-crossing the vertebral spines. What purpose did this arrangement of bones serve?
25. Which two anatomical features in *Edmontosaurus*, a hadrosaur, suggest that it did indeed venture into water?
26. Who named *Hypsilophodon*? Was it (a) T. H. Huxley in 1870; or (b) O. C. Marsh in 1880; or (c) G. Mantell in 1850; or (d) E. D. Cope in 1890?
27. *Ornithomimus asiaticus* has been renamed in a genus of its own. What is this new ornithomimosaur genus?
28. There are at present over fifteen recognised species of *Triceratops*, but, in the future, how is this likely to be reduced?
29. In which country was *Micropachycephalosaurus* found?
30. What does the name *Archaeopteryx* mean?

Quiz 25

1. After 1978 when John Horner discovered the *Maiasaura* nests in Montana, he later discovered the nearby nesting colony of which other dinosaur?
2. Which is the exceptionally large coelophysid or halticosaurid?
3. Which palaeontologist was 'Almost King of Albania'?
4. To which dinosaur family does *Wannanosaurus* belong?
5. Which genera has *mantelli* as a species name: *Iguanodon, Tyrannosaurus, Megalosaurus, Diplodocus* or *Heterodontosaurus*?
6. Which tyrannosaurid was the 'alarming or frightful reptile'?
7. Which marine dinosaur contemporary had 'THUS YOU CHAIRS' as an anagram?

8. Identify this dinosaur:

9. Most casts of *Diplodocus* in the world's museums have been fitted with a wrong part. What is this wrong part and to which dinosaur does it really belong?
10. Which of the following dinosaurs made these footprints: *Tyrannosaurus*, *Brachiosaurus*, *Stegosaurus*, *Iguanodon* or *Coelophysis*?

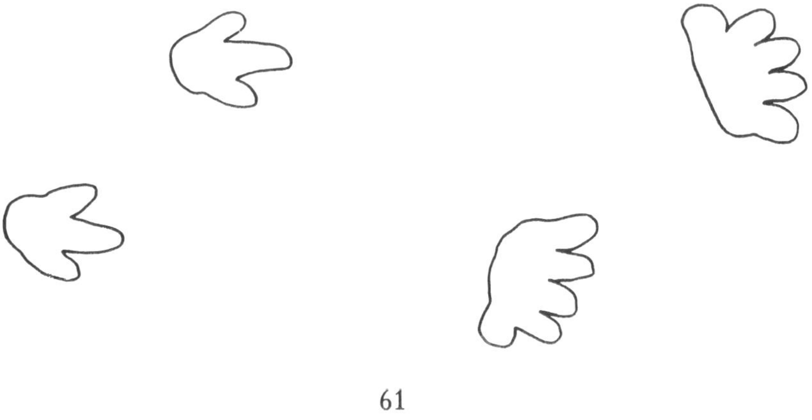

11. Which stegosaurid had 'RIX SUAVE SOUL' as an anagram?
12. Which German palaeontologist could be described as Europe's greatest 'dinosaurologist' of the early twentieth century?
13. Which of the following dinosaurs is the odd-one-out: *Dracopelta, Scelidosaurus, Hylaeosaurus, Sauropelta* or *Nodosaurus*?
14. How many digits did *Centrosaurus* have on (a) its front legs and (b) its back legs?
15. Of all the dinosaurs so far found, which has the biggest claw?
16. What is this?

17. In which country was the recently-described carnosaur *Carnotaurus* found?
18. Most large carnivores kill their prey with large downward-pointing, stabbing teeth. What was different, in this respect, about the teeth of *Deinonychus*?
19. Which Chinese dinosaur was probably very closely related to *Plateosaurus*?
20. How did a plesiosaur use its flippers, like the oars of a rowing boat or up and down like a penguin?
21. How do the pelvic bones differ between crested and non-crested hadrosaurs?
22. Where was *Hypsilophodon* discovered?
23. How many horns did *Chasmosaurus* have?
24. How is the jaw of *Huayangosaurus* different from that of *Stegosaurus*?
25. What was the probable diet and lifestyle of *Pteranodon*?

26. Neave Parker, an artist who painted many famous pictures of dinosaurs for the British Museum (Natural History), produced a portrait of *Hypsilophodon*. What was wrong with it?
27. Here are three stegosaurid tails belonging to *Stegosaurus*, *Kentrosaurus* and *Tuojiangosaurus*. Which is which?

(a)

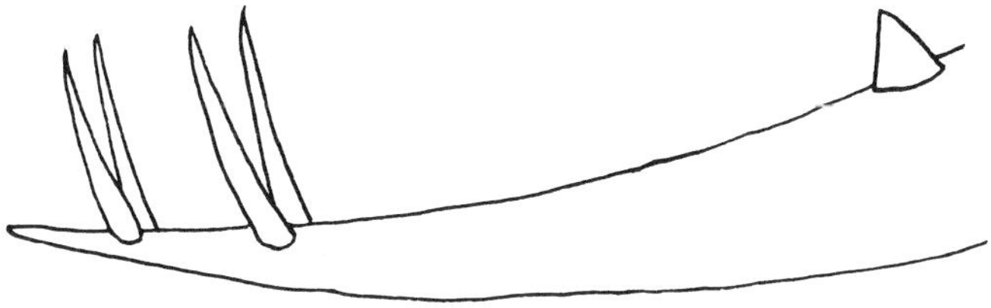

(b)

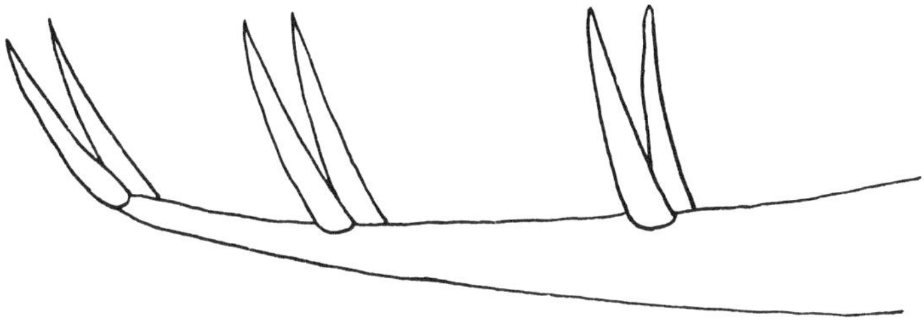

(c)

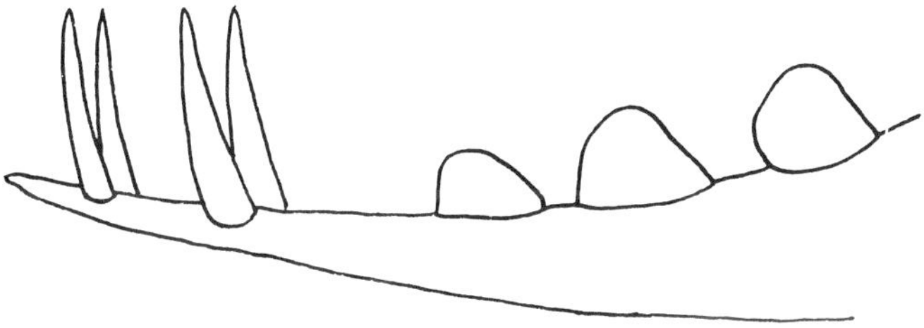

28. Which dinosaur became the first animal cartoon star, in 1914, appearing in a Sunday comic strip 'Little Nemo in Slumberland', in a film and even on stage in a clever variety act with her creator Winsor McCay?
29. What is unusual about the ichthyosaur *Opthalmosaurus*?
30. Which geological period lasted from 225/245 million to 200/205 million years ago?

Quiz 26

1. Identify this dinosaur:

2. Which of the following dinosaurs is the odd-one-out: *Goyocephale, Gravitholus, Yaverlandia, Euoplocephalus* or *Tylocephale*?
3. To which ornithopod family does *Technosaurus* belong?
4. Of which small saurischian dinosaur is 'NASTY RUSS' an anagram?
5. Who supervised (a) the recovery of the Bernissart (Belgium) Iguanodons in 1878-1881 and (b) their scientific description?
6. The Bernissart Iguanodons might be two species. One (*Iguanodon bernissartensis*) is large while the other (*I. mantelli*) is a lot smaller. Two species may have lived side by side but what is a different interpretation?
7. 'The class Dinosauria is proposed by some to include dinosaurs and birds.' True or false?
8. Which of the following dinosaurs was typically 23 metres long: *Compsognathus, Diplodocus, Brachiosaurus, Tyrannosaurus* or *Stegosaurus*?
9. In which two African countries has *Spinosaurus* been found?
10. In sauropods other than brachiosaurids, there are cleft spines in the back and neck vertebrae, producing, along the length of the vertebral column, a canal-like structure. What purpose did this serve?
11. Which dinosaur was found in Lesotho in 1970 but was first called *Fabrosaurus*?

12. Identify this ceratopsian:

13. Identify this dinosaur:

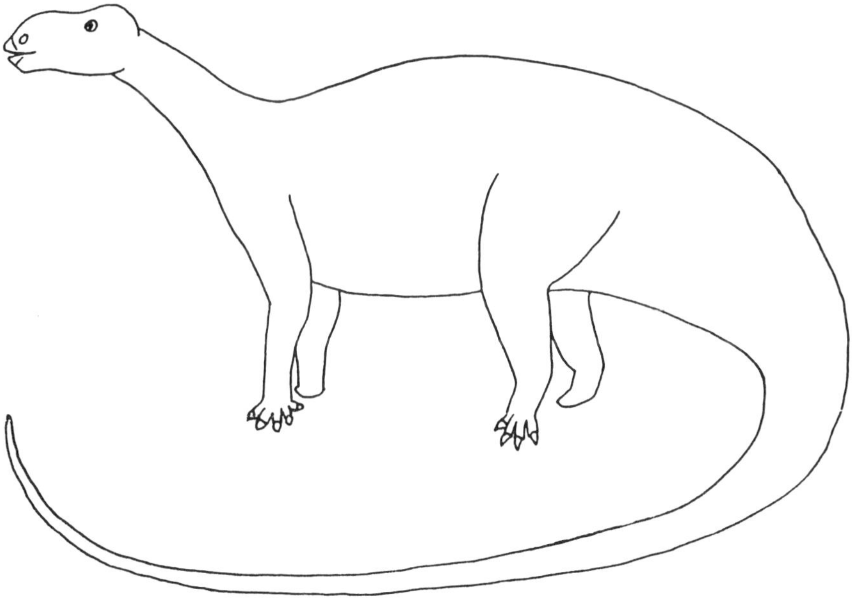

14. *Tenontosaurus, Telmatosaurus, Titanosaurus* and *Tyrannosaurus* are all dinosaurs with similar sounding names. One was a hadrosaur, one a carnosaur, one a sauropod and one a hypsilophodontid. Which is which?
15. What was unusual about the teeth of the pterosaur *Pterodaustro*?
16. Which nodosaurid is usually reconstructed with a row of bony spikes on its flanks?
17. The bony plates of *Scelidosaurus* are arranged separately in rows along its body but where on its body did it have three plates clustered together?
18. Which of these stegosaurs is the odd-one-out: *Stegosaurus, Tuojiangosaurus, Chialingosaurus, Huayangosaurus* or *Wuerhosaurus*?
19. Where have all the remains of *Pentaceratops* been found?
20. How many fingers did *Hypsilophodon* have?
21. Of the fifteen or so species of *Triceratops* (so far recognised) which is the smallest: *Triceratops flabellatus, T. prorsus, T. hatcheri, T. obtusus*, or *T. albertensis*?
22. Of the fifteen or so species of *Triceratops* which is the biggest: *Triceratops serratus, T. elatus, T. calicornis, T. eurycephalus* or *T. horridus*?
23. *Tarchia* is a little-known ankylosaurid from Mongolia. It had a fairly large brain but is known only from its skull. What does its name mean?
24. Who first realised that a fossil found as early as 1784 was a flying reptile?
25. Which of the following dinosaurs possessed this tooth: *Diplodocus, Tyrannosaurus, Stegosaurus, Triceratops* or *Iguanodon*?

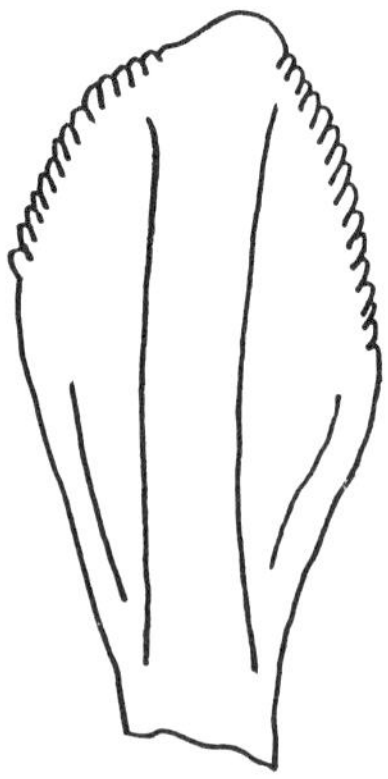

26. What were trilobites?
27. What was the difference between the sails of *Dimetrodon* and *Edaphosaurus*?
28. 'The pterosaur wing was attached to its hind feet.' True or false?
29. Identify this dinosaur from its description: It is a 1.2m long former member of the Fabrosauridae, now placed in the Scelidosauria. It possessed a very long tail, necessary for balancing the body weight, which was increased by the presence of hundreds of small bony studs. It comes from the Early Jurassic and its name means 'small shield reptile'.
30. Which is the most primitive (or earliest-known) hadrosaurid?

Quiz 27

1. Identify this dinosaur:

2. Which of the following dinosaurs was about 80cm long: *Saltopus*, *Compsognathus*, *Cetiosaurus*, *Stegoceras* or *Ornithomimus*?
3. Which bird is 'mimicked' in the new ornithomimosaur *Anserimimus*?
4. Who in 1904 wrote an article about *Brachiosaurus* as a land animal but was ignored or ridiculed until the 1970s when he was finally vindicated?
5. Which two South American palaeontologists discovered *Saltasaurus* in Salta Province, Argentina?

6. Put this classification into the correct order by matching the names against the taxonomic divisions:

Kingdom	Vertebrata
Subkingdom	Eumetazoa
Branch	Reptilia
Grade	Archosauria
Division	Sauropodomorpha
Phylum	Diplodocidae
Subphylum	Animalia
Class	Bilateria
Subclass	*Diplodocus*
Infraclass	Metazoa
Superorder	Diapsida
Order	Saurischia
Suborder	*carnegii*
Infraorder	Deuterostomia
Family	Archosauromorpha
Genus	Chordata
Species	Sauropoda

7. Which dinosaur, known as the Rutland Dinosaur, discovered in 1968 was still 20 years later the most complete dinosaur found in Britain?
8. Where was the mid-Cretaceous land of Beringia?
9. *Stegosaurus* did not have very strong jaws so how did it break up its plant food for digestion?
10. Which of the following dinosaurs is the odd-one-out: *Plateosaurus, Lufengosaurus, Mussaurus, Massospondylus* or *Riojasaurus*?
11. Of which small saurischian dinosaur is 'STOP SAUL' an anagram?
12. Who was the district medical officer who acquired *Archaeopteryx* and sold it to the British Museum for £700?
13. When did the last mass extinction take place before the famous one at the end of the Cretaceous?
14. What kind of an animal was *Megazostrodon*?
15. What is the 'living fossil' lizard that is found on an island near New Zealand?
16. Who is the leading exponent, in recent years, of the idea that the dinosaurs were warm-blooded?

17. To which ornithischian family does *Jaxartosaurus* belong?
18. Which dinosaur family could be distinguished by a shoulder blade with a 'pseudo-acromion process'?
19. After the dinosaur period came the Cenozoic era, 65 million years ago to the present day. It can be divided into epochs. Arrange these epochs in the correct order (oldest first): Eocene, Holocene, Pleistocene, Miocene, Oligocene, Palaeocene and Pliocene.
20. Identify this famous dinosaur person, who, during his career, was known as 'Darwin's Bulldog':

21. Was *Monoclonius* a short- or long-frilled ceratopsian?
22. Very recently three new dinosaurs were discovered in the U.S.A., China and Australia. They were: (a) *Leaellynasaura*, (b) *Drinker*, and (c) *Kunmingosaurus*. In which of the three countries was each one found?
23. Young ankylosaurs had separate bones in their skulls. What happened to these bones as they grew older?
24. How did Mr Peter Dodson of the University of Pennsylvania reduce the number of lambeosaurine hadrosaurs from thirteen to three?
25. Of which Jurassic landmass was Australia a part?
26. Which is the earliest-known iguanodontid?
27. Which African hypsilophodontid (named the 'uncatchable reptile') has now been incorporated into the genus *Dryosaurus*?
28. Which of the following genera has *nigeriensis* as a species name: *Diplodocus, Ouranosaurus, Styracosaurus, Tenontosaurus* or *Titanosaurus*?

29. Identify this hadrosaur:

30. The sauropod *Antarctosaurus* was not found in Antarctica. In which country was it found?

Quiz 28

1. Of which hadrosaur is 'MOUSE RUNS TOAD' an anagram?
2. Which is the odd-one-out of the following dinosaurs: *Lexovisaurus, Ankylosaurus, Pinacosaurus, Euoplocephalus* or *Saichania*?
3. What do the following dinosaurs have in common: *Saltasaurus, Ankylosaurus, Nodosaurus, Scutellosaurus* and *Scelidosaurus*?
4. To which ornithopod family does *Geranosaurus* belong?
5. Identify this dinosaur:

6. Identify this pterosaur:

7. Which dinosaur had the longest-known neck?
8. Which dinosaur contemporary had the longest neck relative to its body size?
9. Which fossil hunter discovered the largest number of new genera of dinosaurs?
10. What is the connection between *Hylaeosaurus* and *Polacanthus* apart from both being Isle of Wight nodosaurids?
11. To which saurischian infraorder does *Barosaurus* belong?
12. Which of the following dinosaurs was typically 15 metres long: *Diplodocus, Parasaurolophus, Tyrannosaurus, Cetiosaurus* or *Stegoceras*?
13. Which dinosaur is often reconstructed with a plume of feathers on the back of its head, although there is no evidence to support the existence of such a structure?
14. Correct this spelling mistake: HIPS SILLY FOOT ON DON.
15. Identify this ceratopsid:

16. What is the Chinese word for 'Dinosaur' or 'terrible dragon'?

17. Which group of modern lizards are related to mosasaurs?
18. 'Pelycosaurs (sail reptiles like *Dimetrodon*) evolved into sail dinosaurs like *Spinosaurus*.' True or false?
19. From which geological period do ankylosaurids date?
20. Who discovered *Chasmosaurus* and where? Was it (a) Joseph Leidy in New Jersey, 1858; (b) Othniel C. Marsh in Wyoming, 1890; (c) Edward D. Cope in New Mexico, 1896; (d) Lawrence M. Lambe in Alberta, 1914; or (e) Dr W. Parks in Colorado, 1924?
21. Which nodosaurid was discovered in Australia?
22. Was *Yaverlandia* a high- or a low-domed pachycephalosaur?
23. Most large dinosaurs have between 4 and 6 vertebrae attached to their pelvis, but how many vertebrae did large ceratopsids have there? About five, seven, ten, twelve or fifteen?
24. Of which mammal-like reptile is 'TIMED DROON' an anagram?
25. Identify this dinosaur from its description: It was between 8.5 and 10 metres long. It was a Cretaceous, North American vegetarian. It was called with good reason an 'abnormal crested reptile'.
26. Which were the three major groups of tetrapods contemporary with the thecodonts (the ancestors of the dinosaurs), during the Triassic?
27. Which of the following dinosaurs typically weighed 2 tonnes: *Saltopus*, *Apatosaurus*, *Tyrannosaurus*, *Compsognathus* or *Euoplocephalus*?
28. Identify this hadrosaur:

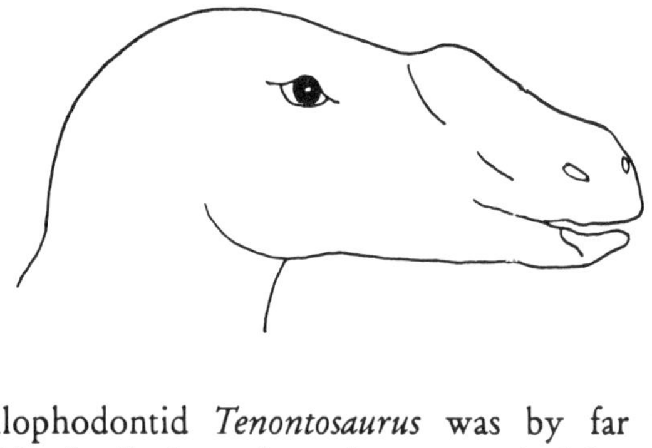

29. The hypsilophodontid *Tenontosaurus* was by far the largest member of its family but what other anatomical feature makes it unusual?
30. What have the following dinosaurs in common: *Antrodemus*, *Apatodon*, *Creosaurus*, *Epanterias*, *Labrosaurus* and *Poicilopleuron*?

Quiz 29

1. Identify this dinosaur:

2. Name the dinosaur which appears in the book 'Jurassic Park' but which is absent from the film, despite having a narcotic bite with which it kills off the billionaire John Hammond.
3. Which of the following dinosaurs weighed typically between 100 and 140 tonnes: *Apatosaurus, Brachiosaurus, Mamenchisaurus, Ultrasaurus* or *Titanosaurus*?
4. What became of *Compsognathus corallestris*, the paddle dinosaur?
5. Which two major groups of dinosaurs appear in the geological record at the same time as flowering plants?
6. Ceratopsians have 'teeth batteries' like hadrosaurs, but what is different about their bites?
7. Which stegosaurid other than *Kentrosaurus* was found in Africa?
8. What is the 'Marginocephalia'?
9. Which nodosaurid was named in 1988 after Charles Sternberg?
10. What is the name of the very close Chinese relative of *Allosaurus*?
11. Of which small saurischian dinosaur is 'TOUGH SPOON CRAMPS' an anagram?
12. Of which mammal-like reptile is 'PROUD AS A HOSE' an anagram?

13. Here are four ornithomimosaurs. One is *Struthiomimus*, one *Gallimimus*, one *Dromiceiomimus* and one *Garudimimus* (which is in a separate family, the Garudimimidae). Which is which?

(a)

(b)

(c)

(d)

14. *Struthiosaurus* is not related at all to *Struthiomimus*. Why not?
15. Name any of the features which place Camarasaurids midway between diplodocids and brachiosaurids.
16. Where and when was *Tyrannosaurus* first found: (a) Colorado, 1900; (b) Texas, 1901; (c) Montana, 1902; (d) Alberta, 1903; or (e) Nevada 1904?
17. Identify this pterosaur:

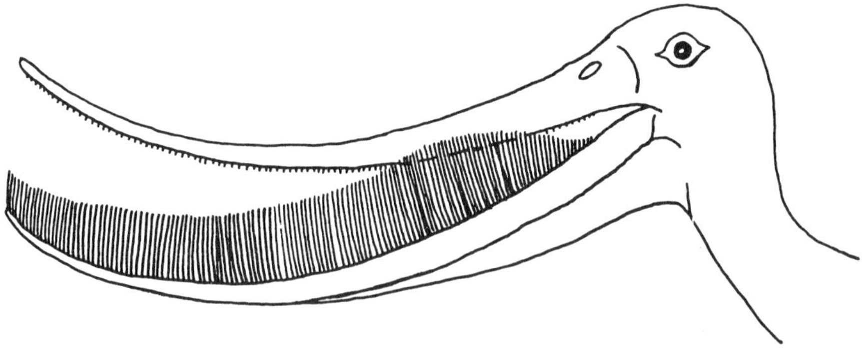

18. To which saurischian infraorder does *Dromaeosaurus* belong?
19. What were the teratosaurs (monster reptiles)?
20. Identify this dinosaur from its description: It was a small, fast-running coelurosaurid from the late Jurassic of North America. It was 2 metres long with small teeth and hands. Very similar to *Coelurus*. Its name means 'bird thief'.
21. Identify this ceratopsid:

22. What was unusual about *Saltasaurus* and *Laplatasaurus*?
23. *Eusthenopteron* was a freshwater fish. What was unusual about it?
24. Which other Brontosaur dinosaur has also, since being named, been shown in fact to be *Apatosaurus*?
25. Which of the following dinosaurs possessed this double-rooted tooth: *Triceratops, Coelophysis, Tyrannosaurus, Stegosaurus* or *Plateosaurus*?

26. *Kuehneosaurus* was a little lizard from the upper Triassic. What did it have in common with the modern lizard *Draco*?

27. What was different about the diets of *Dimetrodon* and *Edaphosaurus*?
28. To which family of dinosaurs does *Panoplosaurus* belong?
29. Which prosauropod family is the 'black mountain reptiles'?
30. In a modern classification of dinosaurs, which groups are included in the ornithischian suborder Thyreophora?

Quiz 30

1. 'The new flowering plants poisoned the dinosaurs.' Why is this an unlikely reason for their extinction?
2. What was the typical wing-span of *Quetzalcoatlus*? Was it 5 metres, 10 metres, 15 metres, 20 metres or 25 metres?
3. Which is the extinct mammal that most closely resembles the ankylosaurs and looks like a huge armadillo?
4. How are ankylosaurid skulls different from those of nodosaurids?
5. What kind of dinosaur was *Tylocephale*, the swollen head?
6. *Pelorosaurus* was named in 1850 but is known only from odd bones, yet it has an amazing number of species. How many species are in the genus and what does the name mean?
7. Identify this dinosaur:

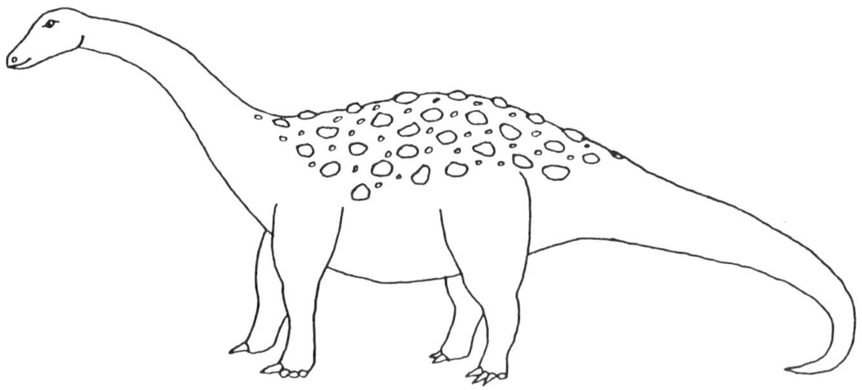

8. Of which prosauropod is 'ACE HOUND RUSTS TOO' an anagram?
9. To which dinosaur family does *Bactrosaurus* belong?

10. Which of the following dinosaurs is the odd-one-out: *Struthiomimus, Gallimimus, Dromiceiomimus, Avimimus* or *Ornithomimus*?
11. What was unusual about the teeth of the pterosaur *Rhamphorhynchus*?
12. In over half the length of the ankylosaurid tail, the vertebrae have ossified tendons. What purpose would these have served?
13. Who, in 1888, showed that the brain of the pterosaur *Scaphognathus* was very bird-like?
14. Which is the only pachycephalosaur found in Britain?
15. Which of the following dinosaurs was typically 18 metres long: *Ceratosaurus, Camptosaurus, Camarasaurus, Chasmosaurus* or *Coelurus*?
16. *Eudimorphodon* probably ate fish. What did *Dimorphodon* probably eat?
17. Identify this pterosaur:

18. Which were the first animals to achieve flight?
19. Which group of Triassic marine reptiles most resembled turtles in their appearance?
20. Lawrence Lambe first called *Euoplocephalus* '*Stereocephalus*' but was obliged to change its name in 1910. Why?
21. Which two nodosaurids survived to the end of the Cretaceous?
22. Was *Micropachycephalosaurus* a high-domed or low-domed pachycephalosaur?
23. Which of the following ceratopids is the odd-one-out: *Chasmosaurus, Torosaurus, Pentaceratops, Styracosaurus* or *Anchiceratops*?

24. To which Jurassic land mass did North America belong?
25. The majority of palaeontologists agree that dinosaurs were warm-blooded, but with an important qualification which sets them apart from modern warm-blooded animals. What is this qualification?
26. What was the probable cause of the mass extinction at the end of the Permian period, 144 million years before the dinosaurs' own mass extinction?
27. What was special about the jaws of dicynodonts?
28. Before the dinosaurs which geological period is known as the 'Age of Fishes'?
29. Which of the following dinosaurs possessed this lower jaw: *Hadrosaurus, Stegoceras, Stegosaurus, Tyrannosaurus* or *Mamenchisaurus*?

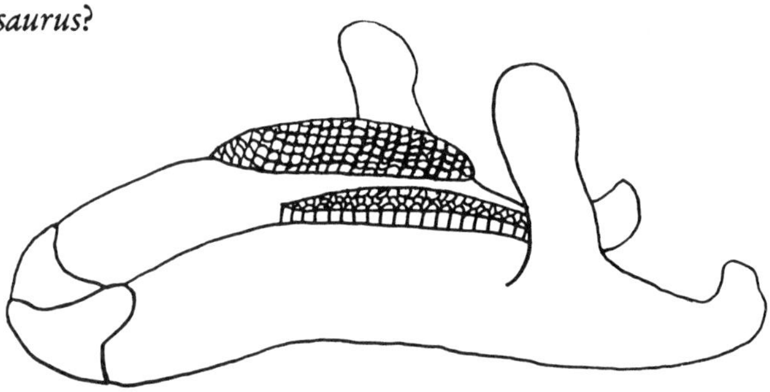

30. '*Quetzalcoatlus* is named after an Aztec God and means feathered serpent.' True or false?

Quiz 31

1. Of which sauropod is 'VAN ON CLOUD' an anagram?
2. To which typical length did the spines of *Spinosaurus* grow: 1m, 2m, 3m, 4m or 5m?
3. Which two dinosaur families are the earliest (and only) representatives of the Ornithopoda at the start of the Jurassic?
4. Which of the following dinosaurs was typically 1.5 metres long: *Cetiosaurus, Styracosaurus, Compsognathus, Psittacosaurus* or *Palaeoscincus*?
5. What was the first (but completely erroneous) scientific name ever given to a dinosaur, by Richard Brookes in 1763?
6. Which geological period could be called the 'Age of Paramammals'?

7. Identify these dinosaurs:

8. Identify this famous dinosaur person:

9. To which ornithischian suborder does *Pachyrhinosaurus* belong?
10. Which of the following dinosaurs is the odd-one-out: *Iguanodon, Brontosaurus, Cumnoria, Polyonax* or *Stereocephalus*?
11. Which was the largest-known ornithomimosaur?
12. Which displayed dinosaur skeleton has been seen by more people than any other?
13. Which dinosaur had the biggest-known hands?
14. Of which marine dinosaur contemporary is 'US YOUR LAST' an anagram?

15. What kind of dinosaur pelvic girdle is this and to what kind of dinosaur would it belong?

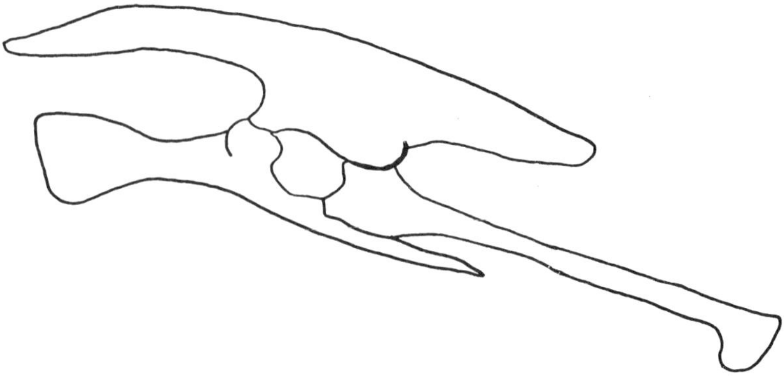

16. Which Australian dinosaur was named after a mythical Greek giant?
17. Which Renaissance man became the first person since the ancient Greeks to say that fossils were all that was left of animals that had lived a long time ago?
18. What kind of an animal was *Mastodon*?
19. What have the following dinosaurs got in common: *Iguanodon, Cetiosaurus, Hylaeosaurus, Megalosaurus* and *Hypsilophodon*?
20. Who wrote 'The Fossils of the South Downs or Illustrations of the Geology of Sussex', published in 1822?
21. 'The Komodo dragon is a relic of the Dinosaur Age.' True or false?
22. What is the connection between *Megalosaurus* and the early mammal *Phascolotherium*?
23. *Staurikosaurus* and *Herrerasaurus* and the new discovery *Eoraptor* are all very early saurischian dinosaurs. Which dinosaur still has the distinction of being the earliest-known ornithischian dinosaur?
24. Name the rarely-mentioned coelophysid from Southern China which was unusual in possessing small brow horns.
25. Which famous dinosaur person lived from 1831 to 1899?
26. 'Bone structure' is a major argument for dinosaurs being warm-blooded because dinosaur bones are similar to those of mammals. What is the 'counter argument'?
27. Which is the one feature that is found in dinosaurs but which is not found in any other reptile?

28. Which English borough has *Iguanodon* on its coat of arms?
29. Which of the following dinosaurs has *ajax* as its species name: *Diplodocus, Apatosaurus, Iguanodon, Stegosaurus* or *Chasmosaurus*?
30. Identify this dinosaur contemporary:

Quiz 32

1. What evidence is there in dinosaur egg shells that at the end of the Cretaceous the dinosaurs were experiencing great stress?
2. If *Micropachycephalosaurus* has the longest name, which two dinosaurs share the distinction of having the second longest name?
3. Identify this dinosaur and say what is wrong with it:

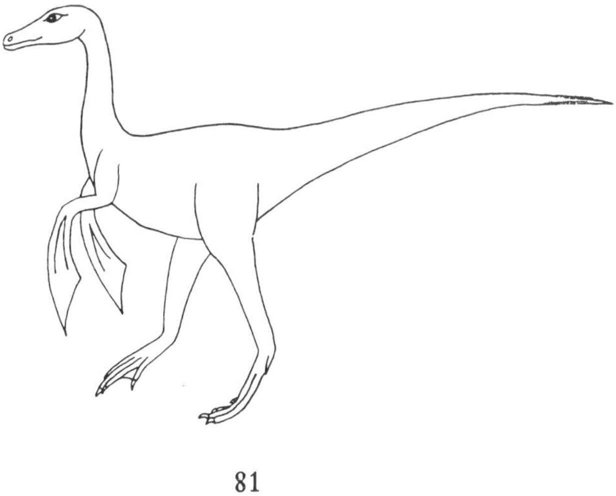

4. Ankylosaurids had S-shaped nasal passages containing thin, curved bones resembling the scroll-shaped bones in the noses of mammals. What function is it reasonable to assume these bones served?
5. Of which sauropod is 'MUM'S SURE A CHAIN' an anagram?
6. Which dinosaur had the biggest eye of any land animal ever?
7. The crests of *Dilophosaurus* were paper-thin. How were they protected from being easily damaged?
8. What was the probable habitat of Pachycephalosaurs?
9. Who decently admitted in 1824 that he had been wrong about *Iguanodon*?
10. What kind of animal was *Euparkeria*?
11. What is unusual about the crest of *Tsintaosaurus*?
12. What was the probable purpose of the cavity in the spinal cord canal in *Stegosaurus* that caused the idea that *Stegosaurus* had two brains?
13. Which dinosaur was named after a famous siege?
14. The family Atlantosauridae is proposed by some authorities. Which dinosaurs would it include?
15. William Buckland is often credited with the naming of *Megalosaurus*. Although Buckland did publish the first description of *Megalosaurus*, who in fact named it?
16. Which of the following dinosaurs possessed this foot: *Camarasaurus, Heterodontosaurus, Tyrannosaurus, Struthiomimus* or *Anatosaurus*?

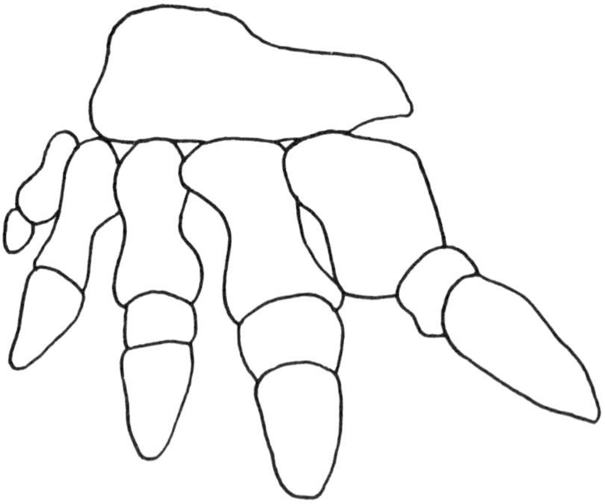

17. Which famous dinosaur person lived from 1769 to 1832?
18. Identify this dinosaur from its description: It was a North American, Cretaceous quadruped measuring 5.5 metres with bands of bony knobs set in its skin. Its name means 'knobbed reptile'. It is in the suborder Ankylosauria.
19. Which dinosaur had the greatest number of neck vertebrae?
20. Where is Como Bluff and why is it important to dinosaur studies?
21. What is the latest known surviving 'paramammal' or mammal-like reptile?
22. Which exhibited dinosaur has a curled tail because there wasn't room in the American Museum of Natural History in New York for it to be straight?
23. What probably caused the contorted position preserved in the remains of many small dinosaurs?
24. Which dinosaur was named after a Lamaist deity?
25. Which of the horned dinosaurs had no horn?
26. Which of the following dinosaurs possessed this tooth: *Iguanodon, Plateosaurus, Stegoceras, Tyrannosaurus* or *Parasaurolophus*?

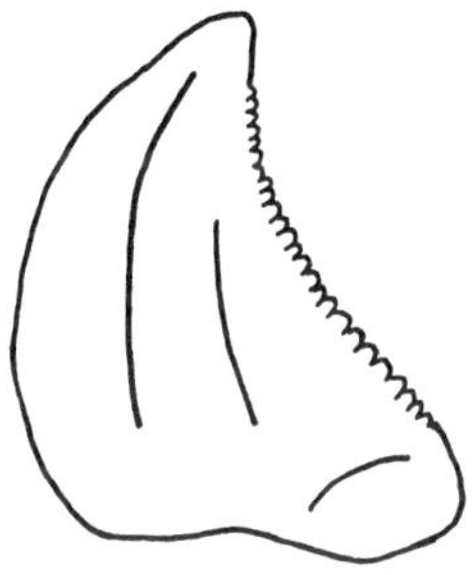

27. What sort of animal was *Platybelodon*?
28. Which group of dinosaurs had their front legs sticking out further than their back legs?
29. Which Argentinean Abelisaurid named in 1966 has a name beginning with the letter 'X'?
30. What are the Deccan Traps and how might they have contributed to the death of the dinosaurs?

Quiz 33

1. Which is the first dinosaur (so far) to appear in the alphabet?

2. Of which ceratopsian dinosaur is 'STOP CAR BY REACH' an anagram?
3. Identify this dinosaur and say what might be wrong with it:

4. For what purpose did the pre-dinosaur archosaurs develop strong back legs?
5. Which of the carnosaur toes points backwards?
6. Which feature of *Archaeopteryx* demonstrates that it could not have evolved from *Compsognathus*?
7. Which is the only known dinosaur that could both swivel one of its toes and rotate its hands on the wrist?
8. Which is the last dinosaur name to appear in the alphabet?
9. Of which group of dinosaur contemporaries was *Shonisaurus* the largest-known member?
10. Name the smallest member of the Coelophysidae.
11. To which dinosaur family does *Silvisaurus* belong?
12. A ceratopsian dinosaur was discovered in 1981 in Montana, but it was not named until 1986, when it was named after the discoverer's wife. Which dinosaur?
13. What is the meaning of the difficult name *Opisthocoelicaudia*?

14. What is the connection between the famous dinner party inside *Iguanodon* in 1853 and the American dinner inside *Mastodon* in 1802?
15. Identify this dinosaur and say what might be wrong with it:

16. Identify this famous Canadian dinosaur person:

17. 'Iguanodontids had large, blunt, hoof-like claws on the fingers.' True or false?
18. Which feature did *Pteranodon* share with the modern pelican?
19. Why, in 1980, was a coelurid-type shin bone from Australia named *Kakuru*, the rainbow serpent?
20. Which fish has been identified from partly-digested remains found within the ribcage of *Baryonyx*?
21. The heterodontosaurid *Dianchungosaurus* discovered recently in China might be a misplaced genus. To which group might it really belong?

22. Which are the three genera belonging to the Family Segnosauridae?
23. What were Noah's Ravens?
24. Which dinosaur had the most rigid tail ever in the animal kingdom?
25. Camarasaurids have been shown to have had a large surface area within their noses. What function would this have served?
26. Which two of the following dinosaurs has *mongoliensis* as a species name: *Saurornithoides, Segnosaurus, Ouranosaurus, Psittacosaurus* or *Nodosaurus*?
27. Which dubious dinosaur, known only from a thigh bone, could be the only European ornithomimosaur?
28. What is the difference between the first toe of ornithomimosaurs and that of coelurids?
29. *Supersaurus* weighed as much as how many large elephants?
30. How is the comparison between *Supersaurus* and so many elephants an effective argument against dinosaurs being warm-blooded?

Quiz 34

1. Identify this dinosaur and say what is wrong with it:

2. How long were the paramammals masters of the Earth before the dinosaurs?
3. *Teraceratops* is not a name given to a dinosaur. What kind of animal was it?
4. By what name is the Chinese Professor Yang Zhong-gian known in the west?
5. Of which hypsilophodontid is 'HAZE YOUR SPURS' an anagram?
6. Which dinosaur has two pairs of back legs?
7. Which is the earliest known dromaeosaurid?
8. In which year was *Archaeopteryx* discovered in Southern Germany?
9. Which other representative of the Nodosauridae, apart from *Polacanthus* and *Hylaeosaurus*, has been found in Britain?
10. Which of the following dinosaurs is the odd-one-out: *Dryptosaurus, Dryosaurus, Coelurus, Nodosaurus* or *Dacentrurus*?
11. To which ornithischian suborder does *Goyocephale* belong?
12. Which Argentinean carnosaur was named in 1979 by Dr J. F. Bonaparte?
13. Which large Argentinean sauropod was named in 1979 by Dr J. F. Bonaparte?
14. Name the South American equivalent of the northern hemisphere's dromaeosaurids. Although not related (since the design of the skull was different), it too had a 'terrible claw'. Its name means North West Argentina reptile.
15. What was the difference between the crests of *Parasaurolophus tubicens* and *Parasaurolophus cyrtocristatus*?
16. There is a dinosaur which is called *Lychorhinus* by some palaeontologists and was also once re-named *Abrictosaurus* by others. What is its third and more accepted name?
17. A recent new theory concerning the feeding habits of *Tyrannosaurus* suggests that it may have hunted using bad breath! How?
18. Name the relative of *Spinosaurus* which is, so far, the only dinosaur found in Thailand?
19. It was proposed in 1914 to rename one of the *Monoclonius* species as *Protorosaurus*, but this was rejected in favour of *Chasmosaurus*. Why was the name *Protorosaurus* rejected?
20. Which hadrosaur had 'straight parts'?

21. Which two new names were proposed in 1916 for *Kentrosaurus* simply because its name was similar to *Centrosaurus*?
22. At the end of the Cretaceous what percentage of living species became extinct?
23. 'Deinonychosaur stiffening rods were the post- and pre-zygapophysis.' True or false?
24. What kind of animal was *Zeuglodon* (also called *Basilosaurus*)?
25. What is most unusual, if true, about the tail of the Chinese sauropod *Shunosaurus*?
26. What is unusual about the short necks of the sauropods *Amargasaurus*, *Rebbachisaurus* and *Dicraeosaurus*?
27. What have the following dinosaurs got in common: *Grallator*, *Amblydactylus*, *Macropodosaurus*, *Masitisisauropodiscus* and *Qomoqomosauropus*?
28. Identify this dinosaur and say what is wrong with it:

29. *Garudimimus* is another of the ornithomimosaurs yet was placed in a family of its own, the Garudimimidae, because it had a small, pointed crest on its skull. What does its name mean?
30. Why did the dinosaurs become extinct?

ANSWERS

Quiz 1

1. Terrible lizard (or reptile).
2. Richard Owen (1804-1892).
3. 1841.
4. None of them are dinosaurs! (a) is *Dimetrodon*, a mammal-like reptile; (b) is a mammoth, a mammal; (c) *Pteranodon* a pterosaur; (d) a plesiosaur and (e) *Glyptodon*, a mammal.
5. Cavemen evolved about 60 million years after the last dinosaur died.
6. *Tyrannosaurus.*
7. *Apatosaurus.*
8. *Diplodocus.*
9. False. Some dinosaurs were quite intelligent and had fairly large brains.
10. (b) Dinosaurs lived only on the land.
11. About 63-65 million years ago.
12. The Mesozoic
13. It would not have been able to inflate its lungs against the pressure of water at that depth.
14. The Hadrosaurs.
15. Pangaea.
16. 4000-6000 million years old.
17. It is usually given as *Compsognathus* at 75-91cm though the true record is held by a baby *Psittacosaurus* at 23-25cm long. The pages of this book are just over 23 cm high.
18. Crocodiles are 'semi-erect' reptiles with their legs sprawling out to the sides and downward, lifting the body clear of the ground, whereas dinosaurs were 'fully-improved' animals, that is, they held their limbs directly underneath their bodies as do modern mammals and birds.
19. (a) Sauropod; (b) Ornithopod; (c) Theropod.
20. Ostrich mimic.
21. A Herbivore.
22. Palaeontology.
23. The Saurischia and the Ornithischia.
24. Birds.
25. The Humboldt Museum in Berlin; *Brachiosaurus.*
26. *Stegosaurus.*
27. True as far as we know.
28. *Triceratops.*
29. *Hypsilophodon.*
30. 23 metres.

Quiz 2

1. To show that the scientific name is being used. In the classification of modern animals we have both the popular name (e.g. domestic cat) and the scientific name (e.g. *Felis catus*). But most dinosaurs have only their scientific name.
2. Both evolved at about the same time.
3. About 150 million years.

4. Seismosaurus (the Earthquake reptile) at 36 metres is at present the longest known dinosaur, longer than *Diplodocus* 27 metres, *Antarctosaurus* 30 metres, *Supersaurus* 30 metres and *Ultrasaurus* 35 metres.
5. *Iguanodon.*
6. (a) Mrs Mary Mantell; (b) her husband Dr Gideon Mantell (1790-1852).
7. *Stegosaurus.*
8. The Ceratopsidae.
9. That they had no internal means of controlling their temperature, which was dependent on the environment.
10. That they could control their temperature by internal biochemical processes, independently of the environment.
11. *Brachiosaurus.*
12. False – they stood the competition for 150 million years.
13. *Tyrannosaurus.*
14. Sedimentary rocks from the Mesozoic era (i.e. rocks formed from sediments like clay, sand or mud).
15. (a) reptile-hipped; (b) bird-hipped.
16. Too big, too slow-moving, too dimwitted and unfeeling, too stupid to respond to change and has lived too long for its own good – none of these insults applies to a dinosaur.
17. Bipedal – two legged; quadrupedal – four legged.
18. The Cretaceous (the last *period* of the Mesozoic *era*).
19. 4,000 million to 4,500 million years ago.
20. The saurischians are half carnivore, half herbivore.
21. A human being, it is believed, but we cannot put it to the test.
22. An animal alive today which was once believed to be extinct and which is known from the fossil record.
23. An extinct prehistoric mammal from the North American Oligocene epoch. It resembled the rhinoceros but was not related to it, other than also being a mammal.
24. *Pachycephalosaurus.*
25. False.
26. Laurasia and Gondwanaland.
27. A row of spines on its back which supported a 'sail' of skin.
28. Because many of their fossilised remains were found in Mongolia in the early 1920s doing just that.
29. *Parasaurolophus* because it is the only vegetarian ornithischian. The others are carnivorous saurischians.
30. *Struthiomimus.*

Quiz 3

1. (a) *Megalosaurus*; (b) William Buckland (1784-1856).
2. Triassic reptiles from which the dinosaurs may have evolved.
3. Definitely false.
4. The Ankylosauridae (or Ankylosaurs).

5. Continental drift.
6. (a) Ectothermic or Poikilothermic; (b) Endothermic or Homoiothermic.
7. *Triceratops.*
8. The 'ruling reptiles', a group including the dinosaurs, crocodiles, pterosaurs and thecodonts.
9. As props to help the animal get upright, to stop its body moving forward while it straightened its back legs.
10. *Brachiosaurus* at 78 tonnes is the usual answer but *Ultrasaurus* weighed an estimated 100 to 140 tonnes, and *Seismosaurus* weighed an estimated 100 tonnes.
11. *Iguanodon.*
12. *Baryonyx walkeri.* This particular specimen remains the only example in the world.
13. An animal that eats fish.
14. That the body is covered quickly by sediment before decomposing.
15. False – *Dimetrodon* was a sail-backed pelycosaur, living before the dinosaurs, in the early Permian.
16. Three.
17. *Stegosaurus.*
18. The sickle-shaped claw on the 2nd toe of the foot.
19. The genus, e.g. in *Tyrannosaurus rex*, *rex* is the species, *Tyrannosaurus* is the genus.
20. Colorado.
21. Baron Georges Cuvier, 1769-1832.
22. An early amphibian of the Permian.
23. Egg thief.
24. The word 'dinosaur' in a loose sense could refer to any large extinct animal. However, if it is to be used as a scientific term, it must be used in the way it has been defined, which is for an animal belonging to one of only two reptilian orders. Ichthyosaurs belong to a different order.
25. Many specimens of *Iguanodon.*
26. *Ankylosaurus* has spines and a bony club at the end of its tail. *Nodosaurus* has neither.
27. (a) Saurischian; (b) Ornithischian.
28. Carnivorous saurischian dinosaurs, usually bipedal, which included the large carnosaurs and the small coelurosaurs.
29. China.
30. It had 3 different kinds of teeth. Its name means 'different tooth reptile'.

Quiz 4

1. *Iguanodon, Megalosaurus* and *Hylaeosaurus.*
2. It separated Gondwanaland from Laurasia, approximately where the Mediterranean Sea is now.
3. *Gallimimus* at 56 km per hour, but possibly its relative *Dromiceiomimus* could reach 65 km per hour.
4. Six.
5. The thumb spike was placed on its nose.

6. False, most were herbivores.
7. In *Stegosaurus* there is a cavity in the spinal cord canal in the area of the sacrum, believed by some to have housed a large nerve ganglion (a small mass of nervous tissue containing many cells) and it is this which some people interpreted as a second brain, but the structure is not in itself unusual and certainly does not amount to being a second brain!
8. The end of a species.
9. An early labyrinthodont amphibian. It was one of the first amphibians.
10. *Brachiosaurus.*
11. *Tyrannosaurus rex.*
12. *Struthiomimus.*
13. Family.
14. A long crest at the back of its head.
15. The pterosaurs.
16. Middle life.
17. The Triassic, Jurassic and Cretaceous periods.
18. *Megalosaurus* – it is the only dinosaur.
19. *Diplodocus.*
20. Benjamin Waterhouse Hawkins.
21. *Stenonychosaurus*, or *Troödon* or its very close relative *Saurornithoides.*
22. When *Compsognathus* was discovered in the 1850s, *Bavarisaurus* had been its last meal (remains were found in the stomach area).
23. *Stegosaurus.*
24. Coal was laid down in the Carboniferous period, long before the dinosaurs.
25. The Ornithomimosaurs.
26. The Pubis, Ileum and Ischium.
27. Ornithischian.
28. Saurischian.
29. *Psittacosaurus.*
30. 9 metres.

Quiz 5

1. Edward Drinker Cope, 1840-1897 and Othniel Charles Marsh, 1831-1899.
2. Saurischian.
3. Ornithischian.
4. Alfred Wegener, 1870-1930.
5. Precambrian, Cambrian, Ordovician, Silurian, Devonian, Carboniferous.
6. *Brachiosaurus.*
7. Woodland reptile.
8. It may have charged at its prey with its jaws agape.
9. *Triceratops.*
10. *Micropachycephalosaurus.*
11. Carnivore.
12. Its thumb spike.
13. 'Double beam'. It refers to the chevron bones of the tail vertebrae which resemble skis.
14. The highly respected Richard Owen believed it to be a whale and nobody questioned it.
15. A giant Devonian placoderm fish.
16. *Deinonychus.*
17. The family.

18. *Stegosaurus* lived in Jurassic North America, *Psittacosaurus* in Cretaceous Mongolia.
19. Ornithischian.
20. Near Vernal in the Uinta mountains, Utah, U.S.A.
21. *Protoceratops*, Mongolia.
22. A suborder of the saurischians including both the sauropods and the prosauropods.
23. The large plates on its back, and the spikes on its tail.
24. False.
25. The subclasses of the class Reptilia. The names refer to the 'windows' in the skulls.
26. Two.
27. 11 metres.
28. Its forehead was massively reinforced with a thick bone structure.
29. Palaeography, which is the study of old handwriting.
30. *Diplodocus*, though some authorities say *Stegosaurus*.

Quiz 6

1. Palaeozoic, Mesozoic, Cenozoic.
2. Permian, Triassic, Jurassic, Cretaceous, Tertiary, Quaternary.
3. Saurischian.
4. *Iguanodon*.
5. The jaws of *Mosasaurus* were discovered.
6. Coprolites.
7. *Deinonychus*.
8. A diagrammatic representation of an evolutionary family tree.
9. Tectonic plates.
10. The ground was littered with so many bones that a shepherd had built a cabin out of them. It is near Medicine Bow in Wyoming, U.S.A.
11. True.
12. A herbivore.
13. *Diplodocus*.
14. *Minmi*.
15. Nearly 2 metres.
16. Ornithischian.
17. Pterosaurs, dinosaurs, phytosaurs, crocodiles, early birds.
18. *Pteranodon*.
19. Because it would not be scientific. It would be like calling a spider an insect. The pterosaur is not classed as a dinosaur because it does not belong in either of the two dinosaur orders. Pterosaurs have their own order because they are so different. The main thing they had in common with dinosaurs is that they lived at the same time.
20. A large prehistoric mammal, not a dinosaur.
21. Cope had described the newly discovered *Elasmosaurus* in 1868 but had mistakenly placed the head on the end of the tail. Marsh pointed this out to him and Cope lost his temper and ended their friendly relations.
22. The replacement of the remains of the animal with mineral – so that the bones become rock. It is one of the processes leading to fossilisation.

23. It was similar in function to the horns of sheep and goats, enabling them to head-butt one another.
24. Sinclair.
25. The blue whale.
26. *Stegosaurus*. The name refers to the extraordinary row of plates along its back.
27. They are older, because they were laid down first.
28. *Ornithomimus* – it is the only saurischian. Also the others have some form of armour.
29. Tree-living.
30. *Staurikosaurus* and *Herrerasaurus* are often given as answers to this question but a new discovery in 1992 in Argentina now qualifies for the distinction of being the earliest known dinosaur, and that is *Eoraptor*, which means 'the thief of the dawn'.

Quiz 7

1. Sir Richard Owen, 1804-1892.
2. False. Birds evolved from a saurischian dinosaur.
3. False. Ornithopods are a suborder of the ornithischia and birds evolved from the saurischia.
4. Saurischian.
5. Travelling down river by boat to locate the samples and then later by a broad-beamed barge, to collect them.
6. *Hypselosaurus* – 30cm long, 2 litre capacity.
7. *Parasaurolophus*.
8. 'Visible life'. From 600 million years ago to the present day, from which there remain recognisable plants and animals (visible). Before this, there survive no recognisable fossil remains (invisible) but there must have been some primitive life forms which were not fossilised.
9. Permineralisation (gradual replacement of the organisms' parts with minerals depositing within). Petrification (the actual remains turning to stone). Natural casts (the complete dissolving and disappearance of the body leaving a mould in the surrounding rock which then fills with mud etc. which then turns to stone).
10. Over a thousand.
11. During the Triassic species could travel freely around the supercontinent of Pangaea but after the breakup of Pangaea into Gondwanaland and Laurasia, in the Jurassic, the oceans prevented further contact and caused species isolation.
12. *Diplodocus*.
13. *Tyrannosaurus rex*.
14. Ornithischian.
15. They are all ceratopsians.
16. *Stegoceras* is the only one that is not a ceratopsian.
17. A large, fierce, prehistoric, ground-living bird from the Palaeocene and Eocene eras of Europe and North America.
18. *Apatosaurus*.
19. A huge pair of hands with big claws.

20. Two ridged reptile.
21. Carnivore.
22. Harry Govier Seeley, 1839-1909.
23. Theropoda and sauropodomorpha.
24. (a) Ichthyosaur; (b) Dolphin; (c) Shark.
25. To hold and tear its prey.
26. The 4th finger.
27. True.
28. To counter-balance the front part of the body.
29. Fossils containing a lot of iron pyrites can easily crumble to pieces. In the early days of fossil collecting, it was considered to be a disease.
30. *Parasaurolophus.*

Quiz 8

1. Thecodonts.
2. False.
3. The part of the cliff above the fossil.
4. Sauropoda.
5. *Ankylosaurus.* The name refers to its armour-plating.
6. *Coelophysis* at Ghost Ranch, New Mexico, 1947.
7. *Styracosaurus.*
8. Ornithomimosauridae and Oviraptorosauridae.
9. *Protoceratops.*
10. Its armoured skin which consisted of alternating bands of small and large nodules of bone.
11. A giant human thigh bone.
12. As a scavenger on the carcasses of dinosaurs killed by other predators.
13. Herbivore.
14. A group of early Triassic archosaurs which looked a little like the modern crocodiles.
15. *Tyrannosaurus* lived in the Cretaceous, *Euparkeria* was a Triassic thecodont.
16. *Megalosaurus.*
17. China in 1958.
18. *Lufengosaurus.*
19. Strictly speaking false; although dinosaurs as a group died out at the end of the Cretaceous, many species and even entire groups had died out long before then.
20. *Dimorphodon.*
21. *Triceratops.*
22. The 1922 American Museum of Natural History expedition to Mongolia which found the famous nests of *Protoceratops.*
23. Baron George Cuvier, 1769-1832.
24. *Saltopus.*
25. 3 fingers and 4 toes (5th toe rudimentary).
26. In the town Dinosaur, Colorado.
27. If an egg is very large it must have a thick shell, but if the shell is too thick the baby dinosaur cannot break out of it. So there is a limit to how big an egg can be.
28. *Euoplocephalus.*
29. High-ridged tooth.

30. It could have helped with feeding, as a tool with which to prod food, but was more likely a defensive weapon.

Quiz 9

1. The horn on its nose.
2. We are referring to a member of the family Ankylosauridae and it is not in italics because it is *considered* to be anglicized.
3. *Spinosaurus.*
4. *Deinonychus.*
5. *Ankylosaurus.*
6. *Diplodocus* was vegetarian.
7. They were a temperature-regulating mechanism.
8. Four.
9. True.
10. *Stegoceras.*
11. *Corythosaurus.*
12. Carnosauria.
13. Taxonomy.
14. *Albertosaurus.*
15. Ceratopia.
16. *Hadrosaurus.*
17. Herbivorous.
18. Mary Ann Mantell.
19. Its name means 'egg thief, fond of ceratopians', and it seems to have been fossilised in a batch of *Protoceratops* eggs.
20. They are all dinosaurs which have been discovered and named in China.
21. The name literally refers to anything that was dug up out of the ground.
22. One of the very first reptiles of the Carboniferous age.
23. *Geranosaurus.*
24. False. Dinosaurs have been found in Antarctica since 1989 (a few armoured dinosaurs and an ornithopod). The freezing conditions make it difficult for us to find them but during the dinosaur times Antarctica was further north and not frozen.
25. *Archaeopteryx.*
26. Three.
27. It had a 'sail' on its back.
28. Thomas Woodrow Wilson, 1856-1924. President from March 1913 to March 1921.
29. It is unrealistic for two reasons: (1) because the mammals had been present throughout the 150 million years of the dinosaurs, but the dinosaurs had still flourished and (2) even if the mammals had eaten the dinosaur eggs, the extinction of the dinosaurs would then have left the mammals foodless, and they would have followed the dinosaurs into extinction.
30.

Class	Reptilia
Subclass	Archosauria
Superorder	Dinosauria
Order	Ornithopoda
Suborder	Stegosauria
Family	Stegosauridae
Genus	*Stegosaurus*

Archosauria is sometimes a subclass, sometimes a superorder, depending on which classification is used.

Quiz 10

1. That Transylvania was an island. Large mainland species often have small island versions.
2. The Pachycephalosaurids.
3. Gazelles.
4. Hadrosauridae.
5. *Torosaurus.*
6. *Baryonyx.*
7. *Tyrannosaurus* was a Cretaceous dinosaur whereas *Brontotherium* was a mammal from the late Eocene.
8. *Ichthyosaurus.* She also discovered the first English pterosaur in 1828.
9. About 215 million years ago.
10. *Ankylosaurus.*
11. The end of the Jurassic. The supercontinent of Pangaea had separated into a northern and a southern continent.
12. The prosauropoda.
13. A small bone at the tip of the lower jaw.
14. True. Theropods are the carnivorous half of the saurischia.
15. True. Tetrapods are four-limbed animals.
16. *Stegosaurus.*
17. *Polacanthus.*
18. *Velociraptor.*
19. A large claw on the first 'finger' of the front legs.
20. For many years most authorities have agreed that the plates were arranged in two alternating rows on either side of the spine. However, recent research and discoveries show that the arrangement must have been a single row down the centre of the back, alternately leaning to each side.
21. *Megalosaurus.*
22. A huge meteorite or perhaps a comet.
23. Pliosaurs.
24. (a) five; (b) four.
25. *Leptoceratops.*
26. *Brachiosaurus.*
27. Probably omnivore.
28. Hen (or chicken) mimic.
29. Gastralia.
30. *Muttaburrasaurus.*

Quiz 11

1. 3,100 million years ago.
2. The size and shape of their crests.
3. Early jawless fish (ostracoderm or shell-skinned fish). One of the earliest vertebrates of the Silurian and Devonian periods.
4. Dean William Buckland, 1784-1856.
5. The Yucatan peninsular, Mexico at a place called Chicxulub (The Devil's Tail).
6. Red.
7. To support the weight of their very large heads.
8. The function of the crests is another controversial dinosaur issue but most authorities now agree that they provided visual signals connected with hadrosaur social behaviour and almost certainly would have been used as 'resonating chambers' to produce strange sounds and calls.

9. When male and female of the same species have different physical forms, such as large frills/crests for males and small frills/crests for females.
10. *Stegoceras*.
11. *Coelophysis*.
12. *Dilophosaurus* is a carnivore – all the others are vegetarian hadrosaurs.
13. In order to lighten the weight of these bones – and to provide extra surfaces for muscle attachment.
14. A row of narrow bony plates.
15. *Apatosaurus*.
16. Three.
17. Psittacosauridae, Protoceratopsidae and Ceratopsidae.
18. Nodosaurs had more defensive armour plating. Ankylosaurs had more offensive weapons such as spikes and clubs.
19. Early Jurassic.
20. A pronounced 'beak', a neck frill and usually horn(s).
21. *Tsintaosaurus*.
22. Prosauropoda.
23. Giraffe-like creatures with long necks feeding from the high branches.
24. True (incredibly).
25. *Dimorphodon*.
26. *Psittacosaurus*, which was 23cm long and is smaller than *Mussaurus* at 25 cm.
27. Herbivore.
28. *Tyrannosaurus*.
29. By hunting in packs.
30. By holding the prey with its hands and disembowelling it with one of its vicious scythe-like claws on its back legs.

Quiz 12

1. *Anatosaurus*.
2. Pachycephalosauridae.
3. To lighten the weight of the bone and to provide areas for muscle attachment.
4. One. The other 'horns' are actually extensions of the frill, not horns.
5. By running away quickly.
6. Four.
7. Kingdom Animalia. Phylum Chordata. Class Reptilia. Subclass Archosauria. Order Saurischia. Suborder Theropoda. Infraorder Carnosauria. Family Tyrannosauridae. Genus *Tyrannosaurus*. Species *rex*.
8. *Pachycephalosaurus*.
9. About seventy.
10. Stomach stones – to aid digestion.
11. They are all famous dinosaur artists.
12. *Microceratops*. Less than 1m long.
13. *Parasaurolophus*.
14. Fabrosauridae.
15. Visual display to establish a social hierarchy.
16. As a temperature-regulating mechanism. It could be used to become active earlier in the day, before its prey. If *Dimetrodon* positioned itself length-ways to the sun, the sail, having a large surface area, would warm up quickly and so warm the animal.
17. A huge Cretaceous turtle.
18. *Rhamphorynchus*.

19. Because sometimes dinosaurs are named from odd fragments of bone, making comparison very difficult, the mistake not being realised until many years later.
20. *Plateosaurus.*
21. Ornithopoda.
22. *Edaphosaurus* – it is a pelycosaur, not a dinosaur, from the early Permian.
23. Laelaps (a hunting dog from ancient Greek mythology) was a name already given to an insect and so it had to be renamed *Dryptosaurus.*
24. *Deinonychus.*
25. Coelacanth (*Latimeria*).
26. A field of comets orbiting the sun at 9 million million miles, believed to strike the Earth every 26 million years.
27. They no longer need them for balance, being quadrupedal.
28. *Deinonychus.*
29. Some believe them to be the same dinosaur. The name *Monoclonius* was given to some fragments of bone by Cope in 1876 while *Centrosaurus* was named by Lawrence Lambe in 1904. The remains of *Centrosaurus* were more diagnostic than those of *Monoclonius* and so *Centrosaurus* should, according to some, be the name adopted. However, both terms are still used, sometimes for the same dinosaur and sometimes not.
30. These four terms *can* be used in an identical sense but, more accurately, the first two refer to the family and the second two to the suborder.

Quiz 13

1. *Pentaceratops* means 'five horned face' but in fact it only had three horns. The supposed 4th and 5th horns were its pointed cheek bones.
2. It would have held it in a rigid straight line to absorb the shock of collision; this was achieved by shock absorbers between the vertebrae which spread the load evenly through the body.
3. 'Pretty jaw'.
4. Carnivore.
5. *Anatosaurus.*
6. The tail vertebrae stiffening rods of *Deinonychus.*
7. To support a temperature-regulating 'sail'.
8. The Hadrosauridae.
9. *Heterodontosaurus.*
10. Prosauropoda.
11. *Trachodon* and *Deinodon*, named by Joseph Leidy, 1823-1897.
12. That they were young *Coelophysis* eaten by cannibalistic adults!
13. *Megalosaurus.*
14. Epoccipitals.
15. Pachycephalosaurs were probably upland, mountain-inhabiting dinosaurs whose remains are rare. After their remains were exposed or washed downstream for some distance, it might be only the skull that remained in a recognisable form. The skull was particularly conducive to fossilisation because it was so thick.

16. The spines in its tail, but some authorities have suggested that it may have flopped onto its side and presented as much armour as possible to the predator.
17. Edward Drinker Cope, 1840-1897.
18. In pairs, in a straight line, so that her feet were always to the sides of the eggs.
19. *Tenontosaurus.*
20. Males have more erect frills and a greater bump on the nose than the females.
21. Sir Arthur Conan Doyle, the creator of Sherlock Holmes.
22. Nodosaurs tended to 'hug the ground' and present armour to the enemy whereas ankylosaurs would rely on their spikes and swing their tail clubs to attack the enemy.
23. Superficially the tail fin looks normal but the bones are only in the lower fin.
24. False since the discovery in 1975 of *Quetzalcoatlus.*
25. Gondwanaland.
26. *Compsognathus.*
27. *Kronosaurus* (a pliosaur).
28. *Ornitholestes.*
29. *Baluchitherium* (also called *Indricotherium*).
30. Dinoturbation is the churning-up and trampling of the ground that can be seen in the fossilised trackways of, for example, a herd of sauropods.

Quiz 14

1. *Zephyrosaurus.*
2. A muscle for lifting the jaw.
3. The Hadrosaurinae (non-crested) and the Lambeosaurinae (crested).
4. A 'hibernation' during the summer to escape from drought.
5. The palpebral bone.
6. *Kentrosaurus.*
7. Ornithopoda in the Ornithischia.
8. The Segnosauridae.
9. *Plateosaurus* is the only prosauropod. The others are sauropods.
10. It was a long-necked plesiosaur. Cope had misplaced the head on the end of the tail and was very angry when Marsh pointed this out.
11. The frill is surrounded at the rim by the normal epoccipitals but the six most posterior terminated in very long pointed spines, forming an impressive visual display.
12. It had no frill.
13. By escaping into water and swimming away.
14. Laurasia.
15. Slim-horned face.
16. *Saurolophus.*
17. *Megalosaurus.*
18. Ceratopsia.
19. Coelurosauria.
20. *Supersaurus.*
21. Gideon Mantell, 1790-1852.
22. Its huge arms.
23. *Tenontosaurus.*
24. Pneumatic openings or air spaces which saved considerable weight.
25. *Bucklandi.*
26. As a means of establishing a social hierarchy amongst themselves.

27. The epoccipital bones surrounding the frill.
28. Foraminiferons, ammonites, belemnites, ichthyosaurs, plesiosaurs, pliosaurs, mosasaurs and pterosaurs.
29. It could be a 'high browser', that is, it could eat the leaves from the tops of trees.
30. By trying to turn them upside down.

Quiz 15

1. Because it could choose between walking on all fours or going bipedally.
2. Herbivore.
3. The 1907 expedition in German East Africa to Tendaguru which produced the famous Berlin *Brachiosaurus*.
4. For balance but also as a 'dynamic stabiliser' for sudden changes of direction at high speed.
5. *Apatosaurus*.
6. *Pentaceratops*.
7. Ornithopoda.
8. *Kentrosaurus*.
9. True.
10. *Coelurus*.
11. It had 3 fingers of which the first was unusually short and the other two unusually long.
12. *Struthiomimus*.
13. *Tyrannosaurus*.
14. By the use of special stiffening rods, called the post- and pre-zygapophyses, attached to powerful muscles at the base of the tail.
15. (a) five; (b) five.
16. It is the famous dinner party organised by Richard Owen and Waterhouse Hawkins for leading palaeontologists of the time inside his model of *Iguanodon*. It took place on New Year's Eve 1853. The famous picture was published in *The Illustrated London News* (January 7, 1854).
17. Because India was an isolated island-continent where species were free of competition from newly-evolved species.
18. The sclerotic ring.
19. *Hypsilophodon*.
20. It was considered *Gorgosaurus* was so similar to *Albertosaurus* that they were the same genus. *Albertosaurus* was named first so the name *Gorgosaurus* became obsolete.
21. Five.
22. Muscular cheeks or cheek pouches.
23. Its name means 'wing finger' and it was so-called because it is the elongated 4th finger which supports the wing.
24. No, they gave birth to live young and we know this because one mother and baby died during the act of birth and became fossilised.
25. *Struthiosaurus*.
26. *Lexovisaurus*.
27. The forward-facing 'tongues' of bone hanging down into the frill window.
28. Dr Jim Hopson of the University of Chicago.

29. Good mother reptile.
30. *Maiasaura* made nests in which the eggs were carefully laid and protected. When the young hatched they continued to be cared for. We know this because in 1978 many nests were found in Montana with young of all sizes.

Quiz 16

1. Othniel Charles Marsh, 1831-1899.
2. *Struthiomimus* had legs that were obviously built for speed, like those of the present-day ostrich. It is reasonable to assume that they had a similar lifestyle.
3. *Tyrannosaurus.*
4. The Cleveland-Lloyd Dinosaur Quarry, Utah.
5. The second toe.
6. *Barosaurus.*
7. Trössingen and Halberstadt in Germany. They have also been found in large quantities at La Chassagne in France.
8. *Quetzalcoatlus.*
9. If *Lesothosaurus* lived in a hot arid climate it would have needed to conserve water and this would have led to a build-up of salts in its body, because salts are normally excreted with water. Instead, the salts would have been excreted through the salt gland, exactly as do some modern reptiles.
10. *Ouranosaurus.*
11. Stegosauria.
12. *Corythosaurus.*
13. *Pteranodon.*
14. Hadrosauridae.
15. If they had been aquatic, the water would have supported much of their weight and they would not have needed such pillar-like legs.
16. They averaged two metres (range between 1.4 and 2.3 metres).
17. Three really – the 4th and 5th horns suggested by its name were really pointed cheekbones, which many ceratopids had.
18. Because it had the characteristics of both groups. A long-frilled ceratopid had a long frill but also long brow horns and a short nasal horn. A short-frilled ceratopid had a short frill with short or no brow horns and a long nasal horn. But *Triceratops* had a short frill with long brow horns and a short nasal horn.
19. Pachycephalosauridae – the head-butting dinosaurs.
20. In the geological record at the Cretaceous and Tertiary boundary there is a thin layer of iridium all over the planet. Iridium is an element rare on earth but common in meteorites, and one theory to explain the death of the dinosaurs is that a large meteorite collided with the earth, releasing iridium into the atmosphere. It eventually settled in an even layer throughout the planet.
21. *Psittacosaurus.*
22. *Dravidosaurus*, India.
23. *Lusitanosaurus.*

24. Bone formed in the skin.
25. *Pterodaustro.*
26. False. There was another mass extinction at the end of the Permian, and there have been many others.
27. *Protoceratops.*
28. The largest known crocodile, found in Cretaceous Texas.
29. Recent life.
30. *Tsintaosaurus.*

Quiz 17

1. *Saichania.*
2. The hadrosaurs – there is evidence for this in the fossilised stomach contents.
3. They are all named after palaeontologists.
4. It is believed that the plates of *Stegosaurus* were a temperature-regulating device and so would need to be richly supplied with blood vessels.
5. *Scelidosaurus.*
6. Absolutely nothing! We can only guess by comparing them to living animals today.
7. They were small with only two clawed fingers, a third being very much reduced.
8. The 'high-domed' (pachycephalosaurids) and the 'low-domed' (homalocephalids).
9. Palaeomagnetism is a recording of the Earth's magnetic field remaining in igneous and metamorphic rocks, as they cool down after their formation. Subsequent movements of tectonic plates result in the palaeomagnetic field shifting out of alignment with the Earth's. Study of palaeomagnetism can be used to trace continental drift and also provides evidence of sudden catastrophic reversals of the Earth's magnetic field, which may be relevent to the interpretation of the fossil record.
10. *Protoceratops.*
11. That it stood on its back legs like a kangaroo.
12. Ornithopoda.
13. True.
14. Mosasaurs.
15. *Iguanodon.*
16. Because it had a horn on its nose, an unusual feature for a carnosaur.
17. *Ouranosaurus.*
18. James Jensen.
19. Their great size.
20. The expansion of flowering plants, the angiosperms.
21. *Monoclonius* or *Centrosaurus.*
22. *Brachiosaurus* is the only large dinosaur, the others are small.
23. A broad toothless beak.
24. The Gobi desert, Mongolia.
25. (a) five; (b) three (4th rudimentary).
26. High-domed.
27. *Iguanodon.*
28. The plesiosaurs had long necks with small heads. The pliosaurs had short necks and large heads.
29. The Jurassic.
30. *Psittacosaurus.*

Quiz 18

1. *Protoceratops.*
2. A 'semi-divided' heart.
3. A fully-divided heart.
4. Winged and toothless.
5. 2.4 metres long.
6. The ceratopids.
7. It was originally thought actually to be a sail, to help it move through the water by catching the wind. The theory is preposterous!
8. *Sarcolestes.*
9. 2.5-2.8 oz (70-80 grams).
10. *Corythosaurus.*
11. Prosauropoda.
12. *Muttaburrasaurus.*
13. 1907, Tendaguru in German East Africa (now Tanzania).
14. *Petrolacosaurus* was the earliest known diapsid reptile from the Carboniferous of North America. Diapsids are a large group comprising modern snakes, lizards, crocodiles and the extinct pterosaurs and all the dinosaurs.
15. False, they were embedded in the skin.
16. It was supposed to have 'paddles' on its arms.
17. Gondwanaland.
18. *Eucentrosaurus.*
19. The solid-crested hadrosaurs.
20. A horn, instead of which was an extraordinary thick, rough plate of bone which some suggest might be a disease symptom rather than natural to the species.
21. (d) *Sarcolestes* (flesh robber) is a misnamed Nodosaurid from England and was a vegetarian. *Hadrosaurus* is also only found in North America.
22. A mosasaur *Tylosaurus.*
23. *Coelophysis.*
24. Ankylosauria.
25. *Brachiosaurus.*
26. A fused collar bone, the furcula or wish-bone.
27. *Parasaurolophus.*
28. *Talarurus.*
29. It meant they could chew their food without it spilling out of the sides of their mouths. This is an advantage because it minimises waste.
30. *Torosaurus.*

Quiz 19

1. In the majority of ceratopian frills (that of *Triceratops* being an exception) there are large openings, or windows, making them rather ineffective as a shield.
2. Dr Anthony R. A. Thulborn.
3. Its name means 'excavated vertebrae', which is mainly all that remains.
4. Europe.
5. *Allosaurus.*
6. The low-domed probably only pushed with their heads, whereas the high-domed pachycephalosaurs head-butted.
7. They are all named after places. Lesotho, Alberta, Dravida (India), Nemegtu (Mongolia) and Itemir (Central Asia).
8. (a) *Tuojiangosaurus*; (b) *Kentrosaurus*; (c) *Stegosaurus.*

9. *Tarbosaurus.*
10. True.
11. *Scelidosaurus.*
12. Ornithopoda.
13. Two paper-thin semi-circular crests on its head.
14. Dr Robert Plot, 1640-1696.
15. *Opisthocoelicaudia.*
16. The Iguanodontid had a thumb spike whereas hadrosaurs had no thumb at all. The Iguanodontid hand was strong for quadrupedal locomotion whereas the hadrosaurid hand was 'paddle-like' and had a slender 5th finger.
17. The lambeosaurine hadrosaurs.
18. North America.
19. The ankylosaurs.
20. The long S-shaped nasal passages and additional sinuses.
21. *Pteranodon.*
22. *Tanystropheus.*
23. Madagascar.
24. An extra attachment area for the jaw-closing muscles.
25. (a) Immediately behind the head on the neck; (b) above the hips.
26. (a) Dr Magdalena Borsuk-Bialynicka.
27. The study of fossilised footprints.
28. Therapsids.
29. Edward Hitchcock.
30. The Squamosal bone.

Quiz 20

1. *Hypsilophodon.*
2. The Cretaceous.
3. Armoured vegetarian Triassic thecodonts superficially resembling phytosaurs.
4. Only high blood pressure, as generated by a fully-divided heart, would enable the blood to reach the head of *Brachiosaurus.* If the necessary extra pressure was to be generated by a semi-divided heart there would be great risk of damage to the circulatory system (e.g. bleeding in the lungs).
5. Today only birds and mammals, both endotherms, have fully-divided hearts; therefore if dinosaurs had a fully-divided heart they ought to be endotherms also. This is fallacious reasoning, though not necessarily untrue.
6. Climatic change over a long period of time.
7. Extinct cephalopod molluscs similar to the modern squid, octopus and nautilus.
8. The frill of *Triceratops* has no openings.
9. That Pterosaurs had fur, contrary to how they are usually drawn which is as leathery creatures.
10. Elephants.
11. Its teeth were like those of sauropods, not prosauropods.
12. Tendaguru in Tanzania.
13. *Nemegtosaurus.*
14. The Hadrosauridae.
15. *Chasmosaurus.*
16. The Hadrosauridae.
17. Long-frilled.
18. *Stegosaurus.*
19. Sauropoda.

20. Ornithomimosauridae. 21. *Pachycephalosaurus.*
22. *Ceratosaurus.*
23. The pile of bones from which Cope identified 3 species came from Ghost Ranch in 1881. In 1947 the same area was excavated and the famous *Coelophysis* graveyard was discovered, showing that Cope had merely recognised the differences in size due to age. There is only one known species of *Coelophysis.* 24. *Chinshakiangosaurus.*
25. The prongs of bone which stick out from the vertebrae and hold them together, as well as providing areas for muscle attachment.
26. It is supposed to have had feathers.
27. The claw of *Baryonyx* was probably on its hand whereas that of *Deinonychus* was on its foot.
28. The Sauropods. 29. About eighty.
30. *Diceratops* lacked a nose horn but apart from that resembles *Triceratops.* The nose horn of *Triceratops* varies greatly and it is probable that *Diceratops* is simply a variant of *Triceratops* and is therefore not considered to be worth mentioning.

Quiz 21

1. Michael Crichton.
2. Only the head of *Nemegtosaurus* is known, and only the body minus the head is known of *Opisthocoelicaudia.* They were found in the same geological formation but until further evidence is found they should continue to be regarded as separate genera.
3. *Yaverlandia,* found on the Isle of Wight.
4. (a) *Lanasaurus* is (4) wool reptile; (b) *Telmatosaurus* is (1) marsh reptile; (c) *Alectrosaurus* is (2) unmarried reptile; (d) *Agathaumas* is (3) marvellous. 5. *Torosaurus.*
6. The natural position is face downwards, presenting the high-domed skull to rivals.
7. Three. 8. *Morganucodon.*
9. Extinction. 10. *Stegosaurus.*
11. (a) When presented towards the sun its surface would have warmed up and the heat transferred to the body, and (b) when angled out of direct sunlight, its surface would have lost heat which would have cooled the body.
12. The Kainozoic. 13. Saurornithoididae in the Deinonychosauria.
14. The name *Troödon* means 'wounding tooth' and it would have been the only meat-eating ornithischian.
15. The fancy neck frill and its tendency to spit venom.
16. (a) Dinosaure = French; (b) Dinosaurio = Spanish; (c) Dinosauro = Italian; (d) Dinosaurier = German.

17. *Lambeosaurus lambei*.
18. Ankylosauria.
19. *Kentrosaurus*.
20. It is an imaginary intelligent 'human-like' being that a dinosaur like *Stenonychosaurus* might have evolved into if it had continued to the present day.
21. The Sternbergs: Charles and his three sons, Charles, Levi and George.
22. Dr Oberndorfer, the lithographic limestone of the Riedenburg-Kelheim area, Bavaria in the 1850s.
23. False; some species are found in North America.
24. Legs adapted to fast running, small heads with jaws full of small sharp teeth, a long flexible neck, and long arms with grasping, sharp-clawed hands.
25. *Diplodocus*.
26. Henry Fairfield Osborn in 1923.
27. Dinosaurs were oviparous, that is egg-laying, like birds. Some snakes are ovoviviparous, that is the embryo develops within an egg but the egg is not laid (it is retained within the mother's body until hatching). Modern mammals are viviparous and so, it is believed, were Ichthyosaurs.
28. Sauropoda.
29. *Triceratops*.
30. The Dashanpu Quarry.

Quiz 22

1. Grasses, because, as a botanical family, they did not evolve until about 25 million years ago, well into the modern age of mammals and well after the last dinosaur had died.
2. *Dilophosaurus* was probably a scavenger of carrion rather than an active hunter and killer of prey. Also, its paper-thin crests would have been easily damaged in any fight. A recent discovery of a new species of *Dilophosaurus* in China suggests that it might even have had a vulture-like beak, giving further evidence for the carrion-eating hypothesis.
3. Fifteen.
4. It had a large claw.
5. *Arrhinoceratops* means 'no nose horn', but it did in fact have a nose horn, albeit small in comparison to the very long brow horns.
6. They all had dorsal spines which may have supported a sail.
7. Charles Robert Darwin, 1809-1882.
8. *Centrosaurus*.
9. *Tsintaosaurus*.
10. The study of how organisms decay and fossilise with a view to understanding why they have been preserved in particular ways.
11. *Diplodocus*.
12. Multiple myeloma, a bone cancer.
13. Segnosauria.
14. Stegosauria.
15. *Dilophosaurus*.
16. The individual bones are welded together.

17. Pleurocoels.
18. *Triceratops alticornis.*
19. *Cetiosaurus.*
20. *Omosaurus* had to be renamed *Dacentrurus* because the name had already been given by Joseph Leidy in 1856 to some other animal.
21. Small openings and three pairs of epoccipitals at the top of the frill, two of which point forward. Also, there are no epoccipitals round the sides of the frill.
22. *Microceratops.*
23. Mr S. Stutchbury of the Royal College of Surgeons.
24. They are now believed to be male and female of the same species.
25. *Thescelosaurus.*
26. The Hadrosaurine hadrosaurs.
27. *Montanoceratops.*
28. A name given to a specimen of *Psittacosaurus*, originally believed to be an ancestral iguanodontid.
29. Fabrosauridae. *Scutellosaurus* is now considered to be in the Scelidosauria.
30. *Tyrannosaurus, Tarbosaurus, Albertosaurus* and *Daspletosaurus.*

Quiz 23

1. *Coelophysis.*
2. Cretaceous.
3. The male had tusks, the female did not.
4. *Mussaurus.*
5. *Megatherium.*
6. *Gallimimus.*
7. *Pentaceratops.*
8. Pachycephalosauria.
9. James Gurney.
10. Hypsilophodontidae.
11. *Brachiosaurus, Dicraeosaurus* and *Kentrosaurus.*
12. When *Iguanodon* closed its jaws the upper jaw on either side moved outwards so that a sideways movement was made against the lower teeth. There are special joints in the upper jaw down both sides enabling this extraordinary movement to be possible.
13. An Oligocene mammal which looked like a rhinoceros with 2 horns but was not related. It has its own order, the embrithopoda.
14. Before the flood.
15. *Lambeosaurus magnicristatus.*
16. (a) They evolved from two or more sources.
17. True. It means slow-moving, lumbering.
18. False, they were. It means fast-running.
19. *Tyrannosaurus.*
20. 136.
21. *Dinichthys.*
22. The name means 'flesh robber'. It was mistaken for a carnivore. Nodosaurs were herbivorous.
23. Stegosauria or Ankylosauria.
24. Short-frilled.
25. Gondwanaland.
26. 1978 near Choteau, Western Montana.
27. *Othnielia.*

28. Hadrosaur teeth were very tough, arranged in dental batteries, hundreds in each jaw, presenting a long, rasping surface. They were well-worn and obviously used to grind up tough plant material like pine needles and wood.
29. *Diplodocus.*
30. Professor Harry Govier Seeley, 1839-1909.

Quiz 24

1. *Oviraptor philoceratops.*
2. *Tarbosaurus* (alarming reptile) is the only one not named after the place it was found.
3. *Ornithomimus.*
4. Ornithopoda.
5. George Peabody.
6. Trace fossils are not the remains of the animals themselves but of traces left by them, such as dung (coprolites), teeth marks, nests and footprints.
7. *Shantungosaurus.*
8. True. They walked on the soles of their feet.
9. True. They walked on their toes.
10. *Vulcanodon.*
11. *Anchiceratops.*
12. *Tyrannosaurus.* A specimen nicknamed Sue had been discovered on an Indian reservation but on a private ranch whose owner had sold it. Ownership of the fossil was in dispute.
13. Its vertebrae were solid with no pleurocoels.
14. Carnosauria.
15. Thirty-nine.
16. *Triceratops.*
17. *Elaphrosaurus.*
18. They are all in the infraorder Deinonychosauria.
19. (b) 33 ft (10 m).
20. Pillar-like hindlegs, small forelegs and a large head on a short neck.
21. *Giraffatitan,* which means 'Gigantic Giraffe', was named in 1988 as a subgenus of *Brachiosaurus.* It comes from Tendaguru in Tanzania where the first *Brachiosaurus* was found.
22. *Barapasaurus,* named from the local (Indian) expression meaning 'big leg'.
23. They are longer than the upper leg bones.
24. The bony tendon trellis held the vertebrae together so that the body was rigid above the pelvic girdle and did not collapse under its own weight.
25. A paddle-like tail, and paddle-like hands.
26. (a) T. H. Huxley in 1870.
27. *Archaeornithomimus.*
28. It is likely that after a detailed study, variations between juveniles and adults and male and female will account for the supposed differences between some species.
29. China.
30. Ancient wing.

Quiz 25

1. The hypsilophodont *Orodromeus.*
2. *Dilophosaurus.*

3. Ferencz (Franz) Baron Nopcsa von Felsö-Szilvas, 1877-1933.
4. Homalocephalidae. *Wannanosaurus* was a primitive bone-head dinosaur described in China in 1977.
5. *Iguanodon.*
6. *Daspletosaurus.*
7. *Ichthyosaurus.*
8. *Homalocephale.*
9. The front feet of the *Diplodocus* on show in the world's museums are the back feet of *Camarasaurus.*
10. *Stegosaurus.*
11. *Lexovisaurus.*
12. Friedrich Freiherr von Huene, 1875-1969.
13. *Scelidosaurus* – all the others are nodosaurids.
14. (a) five; (b) four (with rudimentary fifth toe).
15. *Therizinosaurus.*
16. It may have been part of a *Megalosaurus* thigh bone. It was illustrated in Robert Plot's 1676 book, 'The Natural History of Oxfordshire'. It was believed to be the bone of an elephant or a giant human. In 1763 it was described as '*Scrotum humanum*'.
17. Argentina.
18. The teeth were backward pointed and serrated and thus ideal for tearing large chunks of flesh from its prey.
19. *Lufengosaurus.*
20. Up and down like a penguin.
21. The crested hadrosaurid pelvis is generally of bigger proportions than the non-crested type, with a plate-like extension on the end of the pubis and a hook-like extension on the end of the much thicker ischium.
22. The Isle of Wight.
23. Three.
24. Apart from thicker jaw proportions, *Huayangosaurus* has teeth at the front of the upper jaw, where *Stegosaurus* has none.
25. Fish. *Pteranodon* was almost certainly an ocean-going pterosaur with a lifestyle similar to that of the modern albatross, plucking fish from the surface of the sea.
26. It depicts *Hypsilophodon* as an arboreal dinosaur. It has since been shown that its first toe was not reversed and not therefore suitable to life in the trees. It was clearly a fleet-footed dinosaur.
27. (a) *Tuojiangosaurus*; (b) *Kentrosaurus*; (c) *Stegosaurus.*
28. Gertie, the Wonderfully Trained Dinosaurus (a Brontosaur).
29. As its name 'eye reptile' implies, it had large eyes.
30. The Triassic.

Quiz 26

1. *Polacanthus.*
2. *Euoplocephalus* is an ankylosaurid, the others are pachycephalosaurs.
3. Fabrosauridae.
4. *Syntarsus.*
5. (a) Louis De-Pauw; (b) Louis Dollo.
6. They could be male and female of the same species.
7. True.
8. *Brachiosaurus.*

9. Niger and Egypt.
10. The divided neural spines probably formed a guide for a large ligament acting like a pulley to raise the neck when necessary.
11. *Lesothosaurus.*
12. *Bagaceratops.*
13. *Tenontosaurus.*
14. The hadrosaur is *Telmatosaurus*, the carnosaur is *Tyrannosaurus*, the sauropod is *Titanosaurus* and the hypsilophodontid is *Tenontosaurus.*
15. They were thin, long and numerous and arranged like a comb, to sieve plankton and shrimps.
16. *Hylaeosaurus.*
17. At the back of the head.
18. *Stegosaurus* is the only one found in North America. The others are all from China. Alternatively, *Huayangosaurus* is now placed in its own family.
19. The San Juan Basin, New Mexico.
20. Five.
21. *Triceratops prorsus.*
22. *Triceratops horridus.*
23. Brain.
24. Georges Cuvier, 1769-1832.
25. *Iguanodon.*
26. A class of Arthropods from the Palaeozoic era.
27. The sail of *Dimetrodon* had spines that were straight and smooth. Those of the sail of *Edaphosaurus* were supported by spines that were knobbly.
28. False.
29. *Scutellosaurus.*
30. *Bactrosaurus.*

Quiz 27

1. *Camarasaurus.*
2. *Saltopus.*
3. The Goose.
4. Elmer S. Riggs.
5. José Bonaparte and Jaime Powell.
6. Kingdom Animalia; Subkingdom Metazoa; Branch Eumetazoa; Grade Bilateria; Division Deuterostomia; Phylum Chordata; Subphylum Vertebrata; Class Reptilia; Subclass Diapsida; Infraclass Archosauromorpha; Superorder Archosauria; Order Saurischia; Suborder Sauropodomorpha; Infraorder Sauropoda; Family Diplodocidae; Genus *Diplodocus*; Species *carnegii.*
7. *Cetiosaurus.* It is on display in Leicester Museum.
8. The land bridge connecting North America and Asia where the Bering Straits are now.
9. Stegosaurids probably chewed up their plant food very roughly, swallowed it quickly into a large stomach and left it to ferment. They probably also used gastroliths in the manner of sauropods.
10. They are all large prosauropods except for *Mussaurus* which was tiny.
11. *Saltopus.*
12. Dr Karl Häberlein.
13. At the end of the Permian.
14. A very early Mesozoic mammal.

15. The Tuatura or *Sphenodon*. A reptile that is a very rare living relative of lizards that were abundant during the Triassic. It has changed little in 200 million years.
16. Dr Robert Bakker.
17. Hadrosauridae.
18. Nodosauridae.
19. Palaeocene, Eocene, Oligocene, Miocene, Pliocene, Pleistocene, Holocene.
20. Thomas Henry Huxley, 1825-1895.
21. Short-frilled.
22. (a) Australia; (b) U.S.A.; (c) China.
23. They fused together to form a thick bony covering.
24. He identified ten of them as either male, female or juvenile of the other three.
25. Gondwanaland.
26. *Camptosaurus*.
27. *Dysalotosaurus*.
28. *Ouranosaurus*.
29. *Hypacrosaurus*.
30. Argentina and other South American countries.

Quiz 28

1. *Edmontosaurus*.
2. *Lexovisaurus* is a stegosaurid. The others are ankylosaurids.
3. They all had some kind of bony armour.
4. Heterodontosauridae.
5. *Ouranosaurus*.
6. *Dorygnathus*.
7. *Mamenchisaurus*.
8. *Tanystropheus*.
9. Othniel C. Marsh, 1831-1899.
10. They might be the same dinosaur.
11. Sauropoda.
12. *Tyrannosaurus*.
13. *Syntarsus*.
14. *Hypsilophodon*.
15. *Pachyrhinosaurus*.
16. Konglong or Kunglung.
17. Monitor lizards.
18. False. Pelycosaurs are unrelated to dinosaurs. The sails evolved separately, perhaps for the same reason (temperature control).
19. Cretaceous.
20. (d) Lawrence Lambe, Red Deer River, Alberta, Canada, 1914.
21. *Minmi*. Named after Minmi Crossing near Rosa, Queensland.
22. Low-domed.
23. About ten.
24. *Dimetrodon*.
25. *Parasaurolophus*.
26. Rhynchosaurs, Dicynodonts and Cynodonts.
27. *Euoplocephalus*.
28. *Hadrosaurus*.
29. It had a very long and deep tail, longer than the rest of its body.
30. They are all names that have been given to *Allosaurus*.

Quiz 29

1. *Microceratops*.
2. *Procompsognathus*.
3. *Ultrasaurus*.

4. It was shown to be simply a larger specimen of the *Compsognathus longipes* and there was no real evidence for the 'paddle' interpretation.
5. The Hadrosaurs and the Ceratopids.
6. The teeth cut food to pieces rather than grind it like those of hadrosaurs.
7. *Paranthodon.*
8. The word means 'bordered heads'. It is a modern classification infra-order (or hypoorder) uniting the Pachycephalosauria and Ceratopsia.
9. *Chassternbergia.*
10. *Yangchuanosaurus.*
11. *Procompsognathus.*
12. *Edaphosaurus.*
13. (a) *Struthiomimus*; (b) *Garudimimus*; (c) *Gallimimus*; (d) *Dromiceiomimus.*
14. The similarity of their names is merely coincidental. *Struthiomimus* (ostrich mimic) is an ornithomimosaur, while *Struthiosaurus* (ostrich reptile) is the smallest known nodosaurid, from Eastern Europe.
15. *Camarasaurus* had a short, blunt head with nostrils on top like *Brachiosaurus* whereas *Diplodocus* had a long head, again with nostrils on top. *Diplodocus* and *Camarasaurus* had cleft spines on the neck vertebrae whereas *Brachiosaurus* did not. *Diplodocus* had relatively short forelimbs, *Brachiosaurus* and *Camarasaurus* relatively long. *Diplodocus* had a long neck and tail, *Camarasaurus* and *Brachiosaurus* a relatively shorter neck and tail. *Camarasaurus* also had brachiosaurid type teeth but no 'skids' on the tail vertebrae like *Diplodocus.*
16. (c) Montana, 1902 and described by Henry Fairfield Osborn in 1905.
17. *Pterodaustro.*
18. Deinonychosauria.
19. Early Triassic thecodonts, once thought to be dinosaurs.
20. *Ornitholestes.*
21. *Arrhinoceratops.*
22. They had bony armour plating, unusual for sauropods.
23. It was a rhipidistrian fish, a type considered by some to be ancestral to land-dwelling tetrapods. It had muscular fins and lungs to breath air as well as gills for breathing underwater.
24. *Elosaurus.*
25. *Triceratops.*
26. Enormously long ribs protruding from the sides, which supported a sail of skin, used to allow it to glide from tree to tree.
27. *Dimetrodon* was a carnivore, *Edaphosaurus* a herbivore.
28. Nodosauridae.
29. The Melanorosauridae.
30. The Scelidosaurians, Stegosaurians, Nodosaurids and Ankylosaurids.

Quiz 30

1. Because the dinosaurs existed for millions of years after the flowering plants arrived and seem to have thrived on them.
2. 15 metres.
3. *Glyptodon.*
4. There are four large horns at the back 'corners', and the sides are completely closed by bone, in ankylosaurids. In nodosaurids there are

no horns on the skull and an opening behind the eye. Both have large plates of bones fused together.

5. A high-domed pachycephalosaurid.
6. About twenty. *Pelorosaurus* means 'huge (or monstrous) reptile'.
7. *Saltasaurus.*
8. *Thecodontosaurus.*
9. Hadrosauridae.
10. *Avimimus* is sometimes placed in the ornithomimosauridae but is not like them in many ways. It possibly possessed feathers, and was very small. The others are all standard ornithomimosaurs.
11. The teeth are all long and forward pointing. The tip of the jaw is toothless.
12. For the attachment of the powerful tail-swinging muscles.
13. E. T. Newton.
14. *Yaverlandia.*
15. *Camarasaurus.*
16. Insects and/or small vertebrates.
17. *Dsungaripterus.*
18. Insects.
19. Placodonts.
20. The name had already been used for an insect.
21. *Panoplosaurus* and *Struthiosaurus.*
22. Low-domed.
23. *Styracosaurus* is the only short-frilled ceratopid.
24. Laurasia.
25. That they were warm-blooded not as a result of internal chemical reactions but by virtue of their large size, in other words homeothermic ectotherms.
26. Climatic change due to continental drift and the formation of Pangaea.
27. They probably had a horny bill covering their jaws. Some species had no teeth at all. Some had only a pair of tusks.
28. The Devonian.
29. *Hadrosaurus.*
30. True.

Quiz 31

1. *Vulcanodon.*
2. 2 metres.
3. The Fabrosauridae and Heterodontosauridae.
4. *Psittacosaurus.*
5. *Scrotum humanum.*
6. The Permian.
7. *Stegoceras.*
8. Mary Anning, 1799-1847.
9. Ceratopsia.
10. *Iguanodon* is the only name here that has not been changed.
11. *Gallimimus.*
12. *Diplodocus carnegii.*
13. *Deinocheirus.*
14. *Tylosaurus.*
15. Advanced ornithischian pelvic girdle belonging to dinosaurs like hadrosaurs and ceratopids.
16. *Rhoetosaurus.*
17. Leonardo da Vinci, 1452-1519.
18. Extinct mammal related to the mammoth and elephant.
19. They were all found in Victorian England.

20. Dr Gideon Mantell, 1790-1852.

21. False.

22. They were both found together in the same quarry – so it was known from the beginning of the study of dinosaurs that mammals had existed alongside dinosaurs.

23. *Pisanosaurus.*

24. *Lukousaurus.*

25. Othniel Charles Marsh.

26. Haversian canals, microscopic channels carrying blood and nerves through bones, are present in large mammals and dinosaurs, but it has been shown that this is merely to provide for fast-growing and weight-bearing bones rather than for endothermy.

27. The upright 'fully-improved' posture where the legs are tucked under the body not sprawling out at the sides.

28. Maidstone, Kent.

29. *Apatosaurus.*

30. Ammonite.

Quiz 32

1. When birds are under stress they lay very thin-shelled eggs. Dinosaur eggs found from the very end of the Cretaceous show this sign – some are so thin that the developing embryos would not have been able to absorb enough calcium from the shell to build their skeletons.
2. *Carcharodontosaurus*, *Lancangjiangosaurus* and *Chinshakiangosaurus* all have 19 letters.
3. *Compsognathus corallestis*, the paddle or lagoon dinosaur, reconstructed with paddles instead of hands, on very dubious evidence.
4. To filter, warm and moisten the air and to help provide a good sense of smell.
5. *Mamenchisaurus.*
6. *Dromiceiomimus.*
7. By vertical struts of bone supporting them.
8. Uplands and mountains like the wild sheep and goats of today.
9. Baron Georges Cuvier, 1769-1832.
10. A Thecodont close to the ancestral line of the dinosaurs and possibly also the crocodiles. One of the first reptiles to stand on its back legs.
11. It points forward, it comes from between the eyes and it bifurcates at the tip.
12. It was probably an area containing a glycogen gland, an energy booster for the hind limbs, similar to that seen in birds.
13. *Alamosaurus.*
14. Diplodocids. (Even though some experts do not agree *Atlantosaurus* was a diplodocid.)
15. Dr James Parkinson, who also named Parkinson's Disease.
16. *Camarasaurus.*
17. Baron Georges Cuvier.
18. *Nodosaurus.*
19. *Mamenchisaurus*, nineteen.
20. Wyoming. Many sauropods have been found there, including the renamed *Brontosaurus.*

21. *Oligokyphus.* A water-vole type animal which survived several million years into the age of the dinosaurs.
22. *Tyrannosaurus.* This exhibit was put together from several partial skeletons resulting in a tail that was 3m too long.
23. Contraction of the neck muscles under rigor mortis.
24. *Erlikosaurus* (Erlik was King of the Dead in the Turkik and Mongolian mythologies of central Asia).
25. *Pachyrhinosaurus.* 26. *Stegoceras.*
27. An early elephant. 28. The ceratopids.
29. *Xenotarsosaurus*, the 'strange ankle' reptile.
30. Enormous lava fields in India demonstrating that at the end of the Cretaceous huge volcanic explosions took place in India which were powerful enough to have altered the Earth's climate.

Quiz 33

1. *Abelisaurus.* 2. *Brachyceratops.*
3. *Syntarsus.* There is no evidence that it really had a plume of feathers on the back of its head.
4. For swimming. 5. The first toe.
6. The fact that *Compsognathus*, alone amongst coelurosaurs, has only two fingers, whereas *Archaeopteryx* had three.
7. *Deinonychus.* 8. *Zizhongosaurus.*
9. It was the largest ichthyosaur.
10. The smallest is *Saltopus* at only 60cm. 11. Nodosauridae.
12. *Avaceratops.* 13. Tail bones hollow at the back.
14. A dinner party given by Charles Willson Peale inside his rebuilt skeleton of the American Mastodon predates Waterhouse Hawkins' party by over 50 years. Hawkins' father was present and must have given his son the idea.
15. *Avimimus* – the evidence for the presence of feathers is a little dubious although there were marks on the bones.
16. Charles Hazelius Sternberg, 1850-1943. 17. True.
18. A pouch of skin underneath the lower jaw in which to store fish.
19. *Kakuru* is the first coelurid from Australia, known only from a shin-bone which had turned to the semi-precious stone opal, hence the name 'rainbow serpent'. 20. *Lepidotes.*
21. It might be a prosauropod or even a pachycephalosaur.
22. *Segnosaurus, Erlikosaurus* and *Nanshiungosaurus.*
23. This was the local name to describe a set of mysterious (dinosaur) footprints on a Triassic rock pavement in the Connecticut valley, discovered in 1800 by Pliny Moody, a farm boy from Massachusetts.
24. *Deinonychus.*

25. It was almost certainly a surface area for heat-radiating in order to protect the brain from overheating.
26. *Saurornithoides* and *Psittacosaurus*.
27. *Betasuchus* ('B Crocodile').
28. The first toe, which points backwards in coelurids, is absent in ornithomimosaurs.
29. Fifteen.
30. A large elephant might eat up to 600 pounds (270 kg) of food a day to fuel its endothermy. *Supersaurus*, weighing 15 times the weight of an elephant, despite its smaller surface area to volume ratio, could not have eaten enough food in a day and therefore could not have been endothermic.

Quiz 34

1. Waterhouse Hawkins' 1854 reconstruction of *Iguanodon* as a sprawling quadruped. The thumb spike is placed on its nose as a little horn.
2. Seventy million years.
3. An ophiacodont pelycosaur from the family Eothyrididae.
4. Dr C. C. Young.
5. *Zephyrosaurus*.
6. *Diplodocus carnegii* in the world's museums. The original had missing front legs so replicas of the back legs were placed on the front, complete with three-clawed feet!
7. *Phaedrolosaurus*, the gleaming reptile.
8. 1861.
9. *Sarcolestes*.
10. They are all named by O. C. Marsh and come from North America except *Dacentrurus* from Europe, named by Sir Richard Owen, in 1875.
11. Pachycephalosauria.
12. *Piatnitzkysaurus*.
13. *Patagosaurus*.
14. *Noasaurus*.
15. The crest of *P. tubicens*, like that of *P. walkeri*, is long, gently curving and tubular whereas that of *P. cyrtocristatus* is bent over backwards to point towards the back of the neck. Some claim *P. cyrtocristatus* to be a female of one of the other species.
16. *Heterodontosaurus consors*.
17. The teeth of *Tyrannosaurus* are perfect for building up a particularly nasty and certainly foul-smelling bacterial culture! It is suggested that *Tyrannosaurus* used surprise to take a quick bite out of its prey which then ran away. Over the next few days *Tyrannosaurus* would follow until its prey became so weak with fever and infection that it could no longer escape.
18. *Siamosaurus*.
19. The name *Protorosaurus* had already been given to a Permian lizard-like reptile.
20. *Orthomerus*. The name means 'having straight parts'.
21. *Kentrurosaurus* and *Doryphorosaurus*.
22. 75%.
23. True.
24. A very early whale from the Eocene.

25. It had a tail-club similar to that of an ankylosaurid. This would be unique among sauropods.
26. The spine on top of each vertebra divides, forming a Y-shape. In *Rebbachisaurus* and *Amargasaurus* these were developed to the point of probably supporting a sail with a similar function to that of *Ouranosaurus* or *Spinosaurus*.
27. They are all Ichnogenera, that is, dinosaurs that have been named but are known only from their footprints.
28. *Diplodocus* restored as a 'sprawler'. In order for it to move, there would have had to have been a 'trough' to accommodate the animal's belly. It would have been as dependent on the trough as a train is on its railway lines.
29. Garuda mimic. Garuda was a bird of Indian mythology. *Garudimimus* was found in S.E. Mongolia and named by Rinchen Barsbold in 1981.
30. This just had to be the last question! Your guess could be as good as anybody's, or take your pick from the following (small) selection of possible reasons: Cosmic catastrophe – either a meteorite or a comet colliding with the Earth. Climatic change over a long period of time due to continental drift, sea levels rising or falling etc. The advent of seasons. Sudden climatic change due to non-cosmic environmental disasters such as volcanoes, greenhouse effect, etc. Competition with mammals and birds. Viral epidemics. Act/Will of God. Genetic senescence. All a mistake from the beginning. The 'Dinosauroids' had a nuclear war. Parthenogenesis. Chemical poisoning from a new species of plant first killing the herbivores with the carnivores then starving. Chemical poisoning from increased levels of selenium or nickel due to volcanic or meteorite activity. A sexual imbalance (i.e., too many males or too many females) leading to no further breeding, due to a temperature fluctuation as can happen in modern aligators. Boredom. Aliens took them away. Aliens hunted them to extinction for food or sport. Reversal of the Earth's magnetic field. Atmospheric changes of pressure and/or oxygen content. Gravity changes. They never became extinct: they evolved into birds and flew away. The evolution and unchecked population explosion of moths and butterflies eating all the plants. Nutritional deficiency. Parasites. Becoming too big and stupid leading to pathological problems like bad backs and an inability to adapt and change. Increased solar activity causing blindness before they were old enough to reproduce. Mass suicide due to insanity. They suffocated under enormous piles of their own dung. Just one of those things that happens occasionally and nothing to worry about. They never existed in the first place since their remains were planted by the Christian 'devil' as an illusion.

BIBLIOGRAPHY

The answers to the questions can be found in the following books:

Allaby, Michael, and Lovelock, James, *The Great Extinction* (London, Martin Secker and Warburg, 1983).
Asimov, Isaac, *Dinosaurs* (London, Longman, 1973).
Bakker, Robert, *The Dinosaur Heresies* (Harmondsworth, Penguin, 1986).
Baldwin, Stuart and Halstead, Beverly, *Dinosaur Stamps of the World* (Baldwin's Books, 1991).
Benton, Michael, *Dinosaurs: an A-Z Guide* (London, Kingfisher, 1988).
Benton, Michael, *Prehistoric Animals: an A-Z Guide* (London, Kingfisher, 1989)
Benton, Michael, *The Pocket Book of Dinosaurs* (London, Kingfisher, 1984).
Benton, Michael, *On The Trail of the Dinosaurs* (Crescent Books, 1989).
Benton, Michael, *The Reign of the Reptiles* (Crescent Books, 1990).
Benton, Michael, *The Rise of the Mammals* (Crescent Books, 1991).
Benton, Michael, *Vertebrate Palaeontology* (Harper Collins, 1990).
Benton, Michael, *Dinosaurs* (Godalming, CLB Publishing, 1992).
Carroll, Robert L., *Vertebrate Paleontology and Evolution* (New York, W. H. Freeman and Co., 1988).
Charig, Alan, *A New Look at the Dinosaurs* (London, British Museum [Natural History] 1979).
Colbert, Edwin H., *Men and Dinosaurs* (London, Evans Brothers Ltd., 1968).
Crowther, Peter and Martin, John, *The Rutland Dinosaur, Cetiosaurus* (Leicester, Leicestershire Museums, 1985).
Czerkas, Sylvia J. and Czerkas, Stephen A. *Dinosaurs: A Global View* (Limpsfield, Dragons World, 1990).
Desmond, Adrian J., *The Hot-blooded Dinosaurs: A Revolution in Palaeontology* (London, Blond & Briggs, 1975).
Desmond, Adrian J., *Archetypes and Ancestors: Palaeontology in Victorian London* (London, Frederick Muller, 1982).
Dixon, Dougal, *The Illustrated Dinosaur Encyclopedia* (Octopus Books, 1988).
Donovan, Stephen K. (ed.), *Mass Extinctions: Processes and Evidence* (Belhaven Press, 1989).
Gardom, Tim, and Milner, Angela, *The Natural History Museum Book of Dinosaurs* (London, Virgin Books, 1993).
Glut, Donald F., *The Dinosaur Dictionary*, Bonanza Books, New York, 1984.
Halstead, L. B. and Jenny, *Dinosaurs* (London, Blandford, 1981).
Halstead, L. B., *The Evolution and Ecology of the Dinosaurs* (Wallingford, Peter Lowe, 1975).
Halstead, L. B., *Hunting the Past* (London, Hamish Hamilton, 1982).
Hublin, Jean-Jacques, *The Hamlyn Encyclopedia of Prehistoric Animals* (London, Hamlyn, 1984).

Lambert, David, *Dinosaur Data Book* (The Diagram Group, Facts on File, Oxford, 1990).
Lambert, David, *Collins Guide to Dinosaurs* (Collins, 1983).
Lindsay, William and Fornari, Giuliano *The Great Dinosaur Atlas* (Dorling Kindersley, 1991).
Lockley, Martin, *Tracking Dinosaurs* (Cambridge University Press, 1991).
Lovelock, James and Allaby, Michael, *The Great Extinction* (London, Martin Secker & Warburg, 1983).
Man, John, *The Day of the Dinosaur* (London, Bison Books Limited, 1978).
McLoughlin, John C., *Archosauria, a New Look at the Old Dinosaur* (Harmondsworth, Penguin, 1979).
Michard, Jean-Guy, *The Reign of the Dinosaurs* (New Horizons) translated Thames & Hudson, 1992 (Gallimart, 1989, French original).
Milner, Richard, *The Encyclopedia of Evolution* (Oxford, Facts on File, 1990).
National Geographic Magazine, vol.183, no.1 (January, 1993).
Norman, Dr David, *The Illustrated Encyclopedia of Dinosaurs* (London, Salamander Books, 1985).
Norman, Dr David, *Dinosaur!* (Boxtree, 1991).
Paul, Gregory S., *Predatory Dinosaurs of the World*, Simon and Schuster, New York, 1988.
Reader, John, *The Rise of Life, The First 3.5 Billion Years* (Collins, 1986).
Romer, Alfred S., *Vertebrate Paleontology* (Chicago, University Museums of Chicago, 1933).
Russell, Dale A., *An Odyssey in Time: The Dinosaurs of North America*, (National Museum of Natural Sciences, Canada, 1989).
Stout, William, *The Dinosaurs* (London, Bantam Books, 1981).
Swinton, W. E., *Dinosaurs* (London, British Museum Natural History, 1964).
Thulborn, Tony, *Dinosaur Tracks* (Chapman and Hall, 1990).
Tweedie, Michael, *The World of Dinosaurs* (London, Weidenfeld & Nicholson, 1977).
Wallace, Joseph, *The Complete Book of the Dinosaur* (Gallery Books, 1989).
Wellnhofer, Peter, *The Illustrated Encyclopedia of Pterosaurs* (New York, Crescent Books, 1991).